Published by:
Vasant U. Patel, President
Vitrag Vignan Charitable Research Foundation
HOLISTIC SCIENCE RESEARCH CENTER
Near Mahavideh Teerth Dham,
Kamrej Xing, N.H. 8, Surat, 394185, India
Tel: +91-2621-250750
Email: hsrcsurat@gmail.com
www.holisticscience.org

First Edition: January-2018
Second Edition: December-2018

Editorial Team:
L.D. Patel, Arati Vora, Rajesh Vora

Printed by:
Shree Sai Art & Printers

Design
Isha Vora

© VVCRF 2018

All rights reserved. No part of this publication may be reproduced, stored in a retrieval system, or transmitted, in any form or by any means, electronic, mechanical, photocopying, recording or otherwise, without the prior permission of the Publisher.

ISBN-10: 1727867203
ISBN-13: 9781727867206

SECRETS OF HOLISTIC LIVING
Discovering the power of experiential wisdom

JAYANTILAL S. SHAH

Publisher's Note

It gives me immense pleasure to bring this wonderful book full of experiential wisdom articulated by Shri Jayantilal Shah, Mumbai. The contents given in bunches of sentences are coming from the deep contemplation from the Author's spiritual journey concomitant to his worldly life of over nine decades as an Industrialist businessman.

Any thoughtful person, while pondering over the contents can relate to his own experiences while one endeavour to traverse a spiritual journey especially in the world of business intricacies and daily living full of hubbub of modern day life.

I hope the book will help the readers in their journey towards a higher plane of blissful inner living irrespective of the surroundings. This was the message of Dada Bhagawan that it is possible to live a spiritually fulfilling life backed by correct understanding of Holistic Science of Life and Living.

The author gifted me the manuscript of this book in 2015 for its publication by Holistic Science Research Center. I am happy that his wish for wider reach to the humanity has come true (albeit after his demise) as the book is being launched at an apt time of Plenary Session on 'Philosophy : Theory & Practice' at the great event of Asian Philosophy Conference - 2018 being hosted by HSRC.

Vasantbhai U. Patel
President, Holistic Science Research Center,
VVCRF, Surat
02nd January, 2018

Dedication

This book is dedicated to the soul incarnate Shri Ambalal Muljibhai Patel (1908-1988) popularly known as Akram Vignani Dada Bhagawan who has revealed the inner science of Holistic Life and Living. He was born in the state of Gujarat and lived the life of a businessman in Gujarat and Maharashtra. Since childhood he had a natural quest for scientific explorations and explanations of the human inner life science in a holistic manner encompassing social, religious and spiritual life in day-to-day living, and strived for religious harmony amidst the plethora of heterogeneous belief systems prevailing in the world.

In 1958, he gained his enlightenment while he was sitting on Platform No. 3, at the Surat railway station. Thereafter, he became gradually known amongst people for having the desire to grasp the real scientific understanding of human life. Rather than giving lectures or discourses, he encouraged more and more questions from the seekers of truth and responded thereto.

Dada Bhagawan was a highly evolved person with whom various thinkers, philosophers, scientists and people from all walks of life and religions had dialogues and obtained consistent, logical and scientific explanations to their queries while respecting varied belief systems. In this age of stress, struggle and strain where the requisite harmony of mind, speech and action for inner progress seems disrupted and drained, the knowledge which he revealed comes to the common man as a saviour to attain the goal of happy holistic living with the right understanding of principles and application of holistic science. It helps keeping us equipoise in all circumstances making life sail smoothly.

Holistic Science Research Center, aspires to bring out these truths to the whole world and humanity to enlighten upon the principles behind overall happiness in life & living.

Prologue

"Lessons in Applied Spirituality" originally compiled on 25th May, 1994 (*Buddha -Purnima*), have been found useful for at least hundred people whom I know and many others whom I do not know but are using them by xeroxing the copy.

A time has come for giving a fresh look at the compilation. It was originally meant for encouraging and establishing staff and Directors of Anand Engineers. It did this job well; over and above it helped meditators who came in my contact to get established.

These Lessons form the part of *Chintanmayi Pragna* or wisdom through understanding as opposed to wisdom through spiritual experience, which is the aim of *Vipassana* meditation. The approach of understanding spiritual parameters came from a different origin i.e. our revered teacher who had a influence of Jain tradition and whom we affectionately used to call Dada Bhagwan. I have seen *Vipassana* meditators who got established without reading these lessons, I also found many, who were helped to get established in meditation after being explained these lessons. There are a few of my friends who are neither meditators nor have gone through the lessons and are still practicing these parameters to some extent. However lessons are for those who need them, only difficulty is that they have to be explained. A new lesson has been added on Sanctity of Wealth.

While writing the preface for the previous edition, through oversight I did not mention the names of my daughter Arati and then Secretary Nancy Menezez who were really responsible for the manuscript to see the light of the day.

Many thanks and good wishes to them on behalf of all those who read these lessons.

Jayantilal S. Shah
Tuesday, 1st January, 2002.

Forward

These lessons in applied spirituality are the result of Saturday discourses at the house of Shri Rohit Patel since July, 1991. Over and above Shri & Smt. Patel, two more persons regularly attended the Saturday sessions. Subjects treated were of greater diversity than the lessons presented here.

These discourses were found useful in day to day life of Shri & Smt.Patel at home as well as in the business activity. Smt. Shiloo Patel one of the regular participant suggested that these should be put in writing to facilitate reading and contemplating on its meaning more frequently. This compilation is a result of her request.

Since the beginning of these discourses, it was evident to the participants that the lessons only explain spiritual concepts. To get the results out of these discourses, a meditation practice is necessary and many of us knew only *Vipassana* meditation, this is recommended as an inevitable supplement.

VIPASSANA MEDITATION: This is a rational and transparent meditation practice taught by Rev.Goenkaji in a course of 10 days and above. Practice of *Vipassana* is based on awareness of breathing and body sensations under controlled conditions in a *Vipassana* centre. It results in reduced level of negativities and development of insight in practical matters concerning day to day life. Its ultimate goal is *nibbana* or Liberation from misery & cycles of birth and death. Economic prosperity is a by-product of this practice.

As far as the concepts are concerned I am deeply indebted to my spiritual guide Shri Ambalal Muljibhai Patel, a nearly perfect realized soul of a house-holder who through intimate discourses over a period of five-and-a-half years made me experiment with methods of self-change for achieving successful results in business activity. Over and above the usual functions like management of finance, correspondence, sales, purchase, information statistics, business management deals with management of people in the organization and in the world outside where our operations are involved and it is in this vital but less recognized area that the seed of prosperity lies.

Use of these lessons in business management is based on the hypothesis that prosperity is a direct function of our graceful acceptance of the people as they are, which is the subject matter of Lesson No.1. Other lessons are to help implementation of concepts identified in the first lesson.

What is written in these discourses has already been mentioned by present day saints and holy persons and hence there is nothing exclusive in these lessons for those whose spiritual practices have been developed under the guidance of other spiritual masters of different traditions. These lessons only confirm what they have learnt or at times give new dimension to their understanding of this difficult & subtle subject.

There have been questions about the necessity of a teacher. Although guru or a spiritual guide is a necessity to initiate this process, an effort by the seeker is equally important. During the progress of an aspirant, there are many periods where progress stops, there are frustrations or results are not to the expectations; under these conditions, an individual teacher is very helpful. Although energy level of the teacher has to be higher than the student to make the guidance effective, it has to be pure almost like a laser beam which has infinitely more penetrating power than the electricity from which it is generated.

Presently these lessons are meant for Staff Training Programme of Anand Engineers and for those who feel interested in them.

Jayantilal S. Shah
May 2, 1994 (Buddha Purnima)

Preface

The present book, *Secrets of Holistic Living* by Sri Jayantilal Shah is a welcome addition to the literature on the philosophy of Dadaji, i.e. Sri Ambalal Muljibhai Patel, who has carved a niche for himseslf among the Contemporary thinkers of India. The distinct feature of this work is that it addresses the conspicuous problems that a person faces in performing his roles in daily life situations, as a householder, a worker, a manager , an entrepreneur or carrying out any such responsibility.

Human life is an opportunity for learning that prompts action for progressive evolution towards perfection culminating in self realisation which has been varily termed as *Nirvana, Moksha, Mukti* or 'liberation'. In this journey, there is possibility of one's swerving away from the channel towards the fullfillment without a guide and hence there is a need for a road-map, which is provided by the preceptors from time to time in keeping with the requirements and to suit the contemporary problems. This is exactly what Dadaji's teachings are addressed to and this book is meant to make a person understand the life-circumstances so that he would enjoy every moment of
his existence.

Any person who reads through the teachings of Dadaji as spelt out in this book by Sri Shah, would be struck with admiration at the simplicity and wisdom that is reflected in his speeches and writings. Dadaji brings home a particular point and that is, truth, however complex they may be, can be imparted in a very simple and understandable manner. Sri Shah attempts to give an admiring commentary to the teachings of his teacher by making it focussed in certain particular life issues which all of us encounter in our day-to-day existence.

Dadaji is emphatic about the idea that all religions and ideologies are simultaneously and relatively true since reality is what a person perceives it from his own angle. The real religion is the one that which integrates all aspects of reality, and ultimately leads to self-realisation transcending all possible angularities. This is something that only realized souls can do, since they have visualised the whole in the parts and parts of the whole.

Echoing the profound nature of the world, Dadaji resolutely speaks that even while living in this world and performing all the duties of a householder, it is possible not to lose the 'freedom' even for a second, where nothing touches the tranquility of the core of the 'being' – the Pure Soul. This happens when the man understands that his essential being is transcending body or mind. And he is an instrument functioning as per the *vyavasthita-shakti*, cosmic commendation. Meditation is not something that needs to be done by giving up things and activities but can be performed in and through the things and actions.

For him, both the 'world' and 'pure self (*Brahman*)' are real; however, the world is relatively real and is a temporary adjustment, while *Brahman* is the only the permanent and eternal reality. He says that liberation is the true nature of the soul, and this can only be bestowed upon by a *Gyani*, when a seeker is completely ready for it and goes to the *Gyani* with an open mindset having a desire to know.

The real truth, according to Dadaji as envisaged by Sri Shah is that, when understood, there is no distance between the spiritual life and empirical life. Life is not meant for hoarding things, worrying about himself or hurting others. He brings in the pan-Indian ideal so eloquently spoken in Jainism and i.e., "When attachment, hate and ignorance *(Raga-Dwesha-Agyana)* are removed, liberation is attained". The means to achieve this is by true knowledge, by observing and knowing, being dispassionate and perfectly aloof, anchored in the Pure Soul, with the passing of the thoughts, thus attaining the state of perfect non-attachment. In very expressive way, Sri Shah deals with these concepts for the benefit of those who wish to live to celebrate life in whichever capacity they are or any suffering they are subjected to.

Following Dadaji, Sri Shah attempts to deal with so-called spiritual life by emphasizing that one does not have to renounce the world in order to attain liberation, one has to do away with the ego and attachment, or the "I" and the "Mine". Dadaji gives a very simple solution which is a startling truth and that is: "Fault lies with the sufferer. ". The way to get rid of it is to see one's own culpability, of mis-perception. The problem of self-indulgence, obstinacy, dogmatism, fanaticism, animosity, vengeance and conflicts are beautifully explained by Dadaji, and following him, Sri Shah shows in all these lessons, how in this worldly life of the ocean of sub-atomic particles *(paramanus)*, vibrations are caused due to the veil of ignorance. Sri Shah portrays how happiness and unhappiness are mere mental projections, how memory creates a pull and a tension in the mind, and how comfort and discomfort are mere trappings of the mind.

Dadaji advises seekers to follow *Tri-Yoga* (meditational practices of the three: mind, speech, body) is the beginning of the spiritual pilgrimage, *Gnana-Yoga* (being in the Self-realisation) is the ultimate doctrine, while *Atma-Yoga* (attunement with the real self) is the realisation of one's own identity. He assures them that there is nothing wrong with worldly life, except one's own understanding of it and the dependency that one has created on it. Sri Shah deals with this in a very philosophical way with the theme of Independence and Interdependence.

The book contains twenty-seven lessons dealing with normal ordinary life-situations analyzed in a very simple and profound manner. The themes go around the way a person faces/encounters his life, accepting positively or rejecting negatively. Any negativity is sourced out of one's ignorance and this understanding requires a guide who has realised the world and knows how to impart what he has realised.

Reading through this book is like listening about an experience which is spiritual but not un-worldly. It shows how life is an opportunity to create meaning to enjoy and make others relish the world. This book fulfills every man's deep need to find meaning in this world and when real meaning is discovered it can be spoken of as spiritual. Sri Shah's account is a living narration of Dadaji's thought and I am sure that it will prove to be a valuable contribution to Indian Philosophy during our times. I thank the authorities of HSRC for their admiring effort to bring out this book for the benefit of those who have some acquaintance with Dadaji's teachings and also those who would like to lead a meaningful life by being in the world and not off the world. For further details I would recommend the readers to refer the HSRC book *'Dialogues with Dadaji on Life & Living'*.

Godabarisha Mishra, Ph.D.
Formerly Professor and Head, Department of Philosophy
University of Madras, 14th December, 2017

Contents

1	Eternal Law of Acceptance	01
2	Opinion	05
3	Defence or Safety	08
4	Contemplation on Our Fault	12
5	Charity	15
6	Gratitude	19
7	Forgiveness	22
8	Positive Outlook	25
9	Purification of Desire	29
10	Vision of the World - As it is	33
11	Choiceless Awareness	36
12	Illusion	40
13	Impermanence	44
14	Observation and Pattern Making	48

15	Divine Patterns	52
16	Rhythm & Balance	56
17	Uniqueness of Experience	59
18	Spirituality & Diet	63
19	Independence & Interdependence	66
20	Economic Activity & Competition	69
21	Human Energy	73
22	Husband - Wife Relationship	76
23	Mystery of Death	79
24	Law of Retribution	83
25	Law of Limits	86
26	Sanctity of Wealth	89
27	Human Relations	93
28	Appendix	98

Eternal Law of Acceptance

"What is nature's language? What appears to be justice or injustice by worldly viewpoints is all justice in nature's language. So accept that whatever has happened is just and correct. Even if you are not ready to believe, you will have to accept it after going through agonies." *

~~

This aims at accepting all individuals and situations either with equanimity or gracefully. Any pattern of internal change like meditation, introspection leads to acceptance. In short, acceptance is a technique of internal change to correct external situations like health, business conditions, family problems etc. in a limited way. It is the opposite of reaction.

Response to a situation like sickness with treatment is a well observed practical pattern. Here again, change of life habits, diet etc. that forms the internal change is acceptance. So also in case of business loss. Change of strategy by introspection to correct

---
* All quotes in this book are by Dada Bhagwan

the situation is acceptance. Change of business pattern due to obsolescence is a response, which is the sub-set of acceptance.

The concept of acceptance is at the centre (a middle point of view) of the extreme responses of either non-acceptance or over-acceptance.

Non-acceptance is the (reactionary) effect of over-acceptance; correction needs to be first applied to basis of our non-acceptance by understanding its destructive power.

It is observed that when one does not accept human beings gracefully, whether one likes them or not, the result is loss of wealth and health. This is due to one's failure in managing well his interpersonal relations.

Non-acceptance of things like clothes, furniture and way of living is not as dangerous as non-acceptance of human-beings and situations.

Non-acceptance of human being is usually in case of partners, government officers, family members, neighbours, servants, members of different regional or religious groups etc. As long as somebody is not accepted and hence avoided, men or women worse than those whom we have not accepted are put in our path by nature until ultimately we learn to accept the uniqueness of human beings. So is the case with situations.

It can be learnt by experience and by meaningful observation of different people that non–acceptance usurps lot of energy by way of internal and external friction and thereby reduces the energy level of the human system which may hamper the immunity level making one open to disease.

Whenever we are in a difficult situation, by instinct we try to change the situation whereas by acceptance we overcome our own hostility towards the situation. That is, we ourselves change and the degree of our change is reflected in a change of situation appearing in our favour.

If we cannot change our non-acceptance, then we should gracefully accept the result of our non-acceptance to reduce the damage. This is called secondary acceptance.

Acceptance is a measure of our grace, culture and creativity.
It is cultivated by a strong auto-suggestion to accept. It requires a patient effort over a prolonged duration. It puts life into our system. It is a very important way to come out of our own misery.

When practiced under the guidance of a teacher, it transforms human system into serene divinity enabling intense activity with a calm and steady poise. It cleans the memory lanes by removing HURT spread all over. It is the receding of ego system represented by its manifestations in the form of arrogance, obstinacy, fear, anxiety and restlessness.

Acceptance is not helplessness although usually a situation is accepted when there is no other way. It is the calm acceptance of the people who appear to be responsible for uncomfortable situation by meditating on the real cause of our predicament.

Acceptance is in the ultimate analysis of an inward search for the causes. Taken to an extreme, it is a blessing from all those responsible for our difficulties – an opening of life within and reducing our hostility which over sometime shall be responded by matching positive feelings from them.

Acceptance when properly understood and practiced locates a true spiritual centre line, away from all dualities like love and hate, rigidity and looseness, health and sickness, poverty and wealth.

It is the way to come out of material and spiritual bankruptcy.

If one over-accepts oneself, ego-principle becomes strong and light goes away. If he does not accept himself and starts reducing himself, he becomes a worthless man.

We always get what we deserve: wife, children, business partners, neighbours, colleagues, profit, loss etc. Depending upon the extent of hostility and love within us, some are favourable and some are unfavourable. One can experience that by accepting unfavourable persons gracefully, their number decreases and life becomes more pleasant.

Practice of acceptance over a long period leads to inner silence. It is the essence of non-violence and compassion. It is the aim of all meditation practices.

Of all the non-acceptance, the most powerful one is of death. Strong breath currents built by nature at the time of birth create a strong disbelief in us that we are the body and death of body means our own death. Only slow practice of meditation can take us out of the fear of death. Accepting death gracefully is not to ask for death. It only generates strong life currents within us during our lifetime. There will not be old age.

Acceptance is also the basis of the principle of impermanence (*anitya*), which requires experiencing nature of all sensations as arising – passing away and then calls for their graceful acceptance, be it pleasant or unpleasant.

2

Opinion

"Opinion is the Father of Mind and Language is the Mother of Mind."

~~

Habit of dividing the outside situation in good and bad actually divides the inner psychological energy and slowly reduces it. Acceptance multiplies and opinion divides the energy. Opinion lowers down the level of inner psychological energy.

A human system under the effect of opinion loses the ability to see the fact, especially subtle facts underlying the events.

The loss of psychological energy under the effect of an opinion makes a man/woman, a daydreamer, a wishful thinker, continuously planning (defensively) instead of observing the plans.

Low energy level of the mind opens up the gates of negativity. One tends to become fault finding, pessimistic and restless.

Observation becomes weaker and steadiness is lost. All this happens in the psychological part of the human system.

Opinion results in loss of creativity. There is a constant bubbling of the mind with attendant loss of inner silence.

Opinion results in weakening of nervous system leading to fear psychosis. There is excessive clinging to one's own living.

Stronger the opinion, weaker the mental faculty. It leads to hypersensitivity or loss of sensitivity.

Effort to come out of opinion is equivalent to psychological *pranayama*. It helps in purifying and controlling the mind.

Example:
    1) A man may be arrogant but charitable with a high degree of organizational skill, steadfast determination, able to withstand usual temptations of money. If we judge him by an opinion that he is arrogant and we call him bad if he behaves arrogantely with us, which we do not like, and hence we overlook or take a marginal note of his good qualities.
    2) A politician working under the stress of heavy negative pressures will be judged by his lack of straight forwardness or his acquisition of wealth, which he requires for various purposes to maintain his political career.

Only if we come out of opinion, we will be able to evaluate a man by taking stock of both his negative and positive parameters, without unduly exaggerating his negativities or turning a blind eye towards his positive attributes. Meditation helps to come out of opinion.

Nature keeps on producing dualities of birth and death, growth and decay, ebb and tide, day and night, health and sickness, happiness and misery, male and female. Innumerable dualities create opinion, a love and hate relationship and one gets trapped.

Of all the dualities, the one of pleasure and pain is the most dangerous. In our process of getting rid of a painful experience, the benevolent work of mother nature to take us out of duality is destroyed. Instinct to avoid a painful experience comes from opinion and hence the necessity of the divine knowledge to accept the dualities we come across in the course of life and living.

3

Defence or Safety

"What causes you sorrow? Your feelings of anger, pride, attachment and greed. Why to blame nature for that?"

~~

When some-body makes a mistake of hurting us or harming our possessions through oversight – sometimes even deliberately, we immediately lose our temper, speak harsh words, keep on reminding the concerned party that you made this mistake and so much harm has come to me because of this mistake etc. Such defending action from our side tantamount to offending the other person, who in turn returns to us similar reaction creating the beginning of a vicious cycle.

We should have pro-actively defended ourselves from outcome of the vicious chain of offending and defending by either controlling our temper in the first place or gracefully accepting the happening with equanimity. This is the real defence. It conserves our energy from being wasted on getting entangled in to the vicious cycle. Further, loving and soothing words from us to the other party will protect him or her from adversarial brooding or becoming negative.

At Home: Whenever a house-maid or servant is not coming in time properly, sometimes not attentive to work, firing him might not serve the purpose. There are chances, he will not improve or he will leave the job and we will get somebody worse than him that too after a gap of inconvenience to all members of family. This chain reaction will ultimately make a negative effect upon our homely routine and even health. Best defence in such case is to politely persuade her or him, demonstrating by doing the work ourselves and wait patiently till things improve.

In case, out of old habit, we lose our temper we must apologize during meditation to undo the offence. In that case, all divine faculties will progressively come to our aid to restore health and make life peaceful.

In Business: We try to defend our business by cunningly out-witting a competitor or grab the order by bribing agent of the purchaser or commit execution of work-order within tighter schedules for completing the job earlier in haste. Usually the money earned this way is lost in labour strikes / unrest / sickness (due to overburden) / rejection of goods due to inferior quality, dislocation of the plant / services (due to overload) etc. Best defence is righteousness and wishing well of all our employees and other parties including those who are against us or trying to harm us and one may see that miraculous results will be achieved.

<u>With beggars / person seeking alms:</u> If a man comes to our house asking for food or money, do not doubt his bonafides. Give him something whatever little we can and nature will take care of the rest. This is our defence.

There are situations like being persuaded by beggars / other persons seeking alms, where the defence is to silently bow before the divinity within him. This will prevent the feeling of resentment emerging in us. Payment to the beggar is not very important per se, if you are unwilling to donate. Defence here is to salute the divinity within him and if at all we give something or pay some money, we should do it with reverence and gratitude. In nature, there is no single sided transaction, hence be mindful that blessed we are since he gave us a chance to exercise our charity and feel moments of gracefulness and fulfilment.

<u>Noise:</u> If the situation outside is noisy, as it is during festivals / processions like *Diwali*, do not get irritated; just think of the pleasure others are having. Slowly you will find that noise does not disturb you; you are only aware of the noise and silence. This is your personal defence.

<u>Dealing with naughtiness of Children:</u> It is their birth right to make the things topsy-turvy, play with gadgets with their curiosity, disturb us when we are working, sometimes stop us from working by asking us to play with them etc. Know and realise for sure that you are over-working, your nerves are tense and child is the God's representative to give you a signal to take a pause and relax. If this signal is not heard, doctor will give us bed-rest for two or more weeks!

<u>With Spouse:</u> Best defence against our spouse is never to offend her or him. True love is the most difficult thing to develop. What we ordinarily understand as loving our wife or husband is in fact

an attachment or commitment. Real love consists of accepting the spouse as it is without argument or reason and slowly persuade her or him to do certain things we believe correct. Be aware that many of the times what we believe to be correct is not correct.

Although defending ourselves is needed in life, many a times what actually happens is offending others which initiates a chain of boomeranging. Due to an illusory belief, it appears as if we were defending ourselves but unmindfully it triggers offensiveness in others and generates a cyclical chain of pains. Wisdom lies in developing equanimity as an effective defence mechanism in the long run.

4

Contemplation on Our Fault

"When one starts realizing his own mistakes, there would be progress. One is judge, lawyer and culprit all by himself then what would be a judgment?"

~~

Instinctively it is easy to exactly locate where someone else is going wrong but very difficult to do a similar exercise with one's own self. However, precisely that is what is required in applied spirituality.

Whenever we become a victim of an accident, a business loss, cheating, theft at home or at office or any unpleasant incident whatsoever, there is always someone whom we can and do blame. Although appearing real, in fact it is an illusion. Real reason lie within us. Our too much attachment to anything – money, body, wealth, our self-centeredness using unscrupulous means to achieve something with the exclusion of others, results in loss of wealth, cheating, injury and accident either to us or other family assets.

There are certain cases like a chronic disease, prolonged ailments, a mental depression, cancer or any other diseases, which result in loss of wealth or premature death or both. There is no apparent external reason. In such a case, usually stress caused by excessive work or desire for earning too much of hubbub or travelling, knowing too many things for multi-tasking, eating various tasty foods indiscriminately, materialistic obsession, etc.; all these result in restlessness of the mind. Slowly control over body movement, speech etc. is lost; mind gets lost with too many imaginations. At times, this results in loss of sleep, requiring tranquillizers. Ultimately the system becomes vulnerable to disease. The evidence once again is lying inside.

Aim of a true spiritual practice is not to stop people from earning wealth but to keep these desires and resulting stresses within permissible limits; regular meditation helps to achieve this level. In case this limit is crossed, correction can be applied by introspective meditation so as to resolve to return to normality. Change in lifestyle: change of routine, change of place, good reading, nice company, etc. is also useful to come back to normal from excessive stress.

Many times, people insult us, have rude behaviour towards us, create obstruction in our working by instigating people against us, strikes, lock-outs, gherao etc.; all this is rooted to our arrogance displayed to subordinates in general. These insults and obstructions are to be sensed in a true spirit as a reaction to the inherent negative cause within us to be mitigated by a positive resolve of taking care of the ego of others in our interactions.

Women and men when harbour strong under-currents by way of stored hurts, jealousy or vanity, they come across various difficulties in their marital life because of these negativities and become victims of health problems. If they can find these underlying evidences and make strong auto-suggestions to counter-act them with positivity, they can come out of their difficulty.

In short, no punishment comes to us without guilt on our side, no rewards without earning. Coming out of our negativities bring wealth as a corollary. Searching for the fault within and correcting the same is called the effort. All other efforts are in a way merely movements of mind, speech and body; much of those tend to be negative, unwanted and excessive if the underlying faults within us continue to be preserved as such while keeping busy with attempting redressal of mere symptoms outside.

Living in close contact with the enlightened teacher is initially very important. Following and obeying his instructions in our living, prove to be of a great help in noticing the grave faults underlying within us. Once we notice and accept our own fault as a crucial evidence to generation of all the maladies we are facing, half the remedy is done as in case of a true diagnosis. The other half battle thereafter becomes easier to win as we experience the life this way, find out the deep rooted fault within and replace it with right knowledge. This will steadily elevate the spirit of life.

5

Charity

"Charity is the investment in the safest bank with highest interest."

~~

All householders in all faiths have an obligation to part with a definite portion of earning for helping the orphans, poor people, education, medical relief, spiritual activities etc.

The percentage varies from 5 to 25 of the income. Whereas 25% is the maximum, 5 to 10% is very common.

Isha - Upanishad mentions that all earning must be used after first giving away the part. *Bhagavad-Gita* says that he is a thief who does not give charity and uses everything he earns for his personal use. Dada Bhagawan says that any monetary wealth does not last for more than a time-cycle of eleven years, it is bound to reshuffle and change possession anyway even if one clings too tight to money by not yielding to opportunity for charity. He says charity is an investment in the safest bank (nature) with highest interest, hence is bound to return to the origin in one way or the other.

As far as possible, charity should be done in such a way that it is not publicized. It may really not be a high secret but should not be announced.

A man who does not use part of the income for charity does considerable harm to himself. Flow of divine guidance within him gets blocked gradually and he tends to indulge in addictions, drink, amour and other derelictions. He starts losing wealth by way of family sickness, quarrels, litigations, falling victim of cheating etc.

In modern times, it is seen that uncharitable businessmen lose wealth by staff unrest, labour strikes, lockouts, recession and many other unforeseen calamities.

Charity makes a man pure, benevolent and virtuous. Divine energy starts flowing into his internal system. Charity, which is given for name and fame although being good, leads to arrogance. It is in a way better than no charity yet quite inferior to charity without seeking name and fame.

It is very difficult to remain charitable and humble at the same time but it is possible to move in that direction with conscious effort. Charity is repayment of debt to the nature who provides us air and water, the two most important elements as an act of charity. The only aim of charity should be self-purification.
The Bible mentions that let your left hand not know what your

right hand does – as the law of charity. Over and above the charity of money, benevolence and kindness to our staff, workers etc. is also considered as charity.

When we give charity, we should not think of hundreds of poor people whom we are unable to help. This is a useless exercise and leads to a point of no-return. We are not the only people in the world to do charity. Nature keeps on producing hundreds of needy people and enough rich to make up the balance.

Highest charity is to teach and make people practice benevolence, spirituality, self-knowledge etc. There are a few people capable to undertake the task of imparting the virtues of benevolence, universal brotherhood and spiritual elevation.

Guiding people on the spiritual path, setting up of *ashrams* and meditation centres etc. are also acts of subtle charity although they are dependent upon gross charity.

Saint Poet Tulsidasji has composed a popular couplet which when translated in English reads like this: "Oh Tulsi, river water does not decrease because of drinking by birds."; in a similar fashion wealth does not decrease because it is given as charity; on the contrary it increases by divine help.

Charity should not be done through emotion or impulse but with a conscious feeling of oneness. Nor we should ask others to contribute towards a noble cause, because it can lead to reciprocal liability for us.

Silent charity is a short cut to prosperity by activating divine forces to work for us.

Dada Bhagawan says "Man is a Guest of Nature". Storing money with a distrust of nature is opposite of charity.

Charity without calculation is extravagance.

Charity has the inbuilt component of the graceful acceptance of the world 'as it is'.

6

Gratitude

"One should never forget obligations of Mother, Father and Guru."

~~

Our present way of life which consists of piped water, covered drainage, smoke-free cooking, refrigerators, air-conditioners, telephones, TV, car, office automation, internet, systems of education–everything has originated from west; hence we must be grateful to the western nations and the scientists for having made our life so comfortable. Whenever we get time, we must pray for their well-being. Our goodwill must go to them. We should even be grateful to the British people for the gift of an international language we can use in our communication, system of land revenue, penal code, railways, police, judiciary etc.

We must express our gratitude to parents for looking after us during our childhood, our education, guiding us during our difficult times – in short, playing a crucial role during first twenty years of our life. Whenever we are alone, we must contemplate upon various episodes in life when parents played an important role in changing the course of events. One can refer early chapters of *Autobiography of a Yogi* or life story of any great personality to see how one has remembered the parents with gratitude.

Gratefulness should also be expressed by remembering those episodes in life where certain friends or acquaintances became useful at critical moments.

Gratitude should also be expressed to wife for protecting us during our times of frustration or preventing us from doing certain acts, which would have led to a disaster. Such actions should be remembered with effort.

Gratefulness to the land of our birth however backward and ugly it may be. Gratefulness to all the poor people who give us a chance to exercise charity and allowing us feel worthy. Just imagine a life with no opportunity to do any act of charity, it leads to a feeling of worthlessness.

Gratitude to nature for providing us free air and water, various types of foods and fruits, its beauty of sunrise and sunsets, hills and valleys, flora and fauna, vast expanse of sky and sea, beauty of her forests and falls – the great mother in all her diversity and variety gives us a chance to observe and enjoy her majesty.

Gratefulness to the incidents of sickness, loss of wealth or any other painful experience which became instrumental in changing the course of life.

Namaskar, which is very common to Indian practice is an act of respect and gratitude to all those godly persons who help us to experience a higher level of our consciousness. Gratitude has to be cultivated with prolonged conscious effort to make it meaningful; otherwise it becomes ritual like 'Thank you - Sorry' etc.

Gratitude is one of the principal parameters of applied spirituality. Solitude or retreat is essential for contemplation and developing a spirit of gratitude. No company or too much company may turn counter-productive.

When we go to bed every night, we must recollect something positive that has happened during the day, a pleasant chat with wife and children, a good customer, a loving smile from some holy person, nice music, a good order, a payment which was badly expected – any sweet experience during the day and offer our thanksgiving gratitude for this incident either to the person or to nature or God.

A person who cannot recollect any sweet incident during the day has not developed worthwhile sensitivity to the workings of nature. Every day, there is sweetness except for few days when the painful development process is on, which is in a way similar to the pains of childbirth where the result is sweet.

Gratefulness cannot be experienced by those who have not developed depth and seriousness in life. Consciousness cannot be purified without adequate practice of gratitude.

7

Forgiveness

"Forgiveness does not have to be expressed in words. It is manifested spontaneously. The very sign of self-realisation is natural/innate forgiveness."

∼∼

Everyone is aware of the fact that we store hurt for many years. If someone has not helped us whom we helped, someone told us harsh words many years ago, someone was responsible for our loss of wealth, someone misbehaved with us, came in our way which we considered right, held views or beliefs different from ours and we quarrelled with each other, all these hurts unless cleaned with conscious effort get stored in our memory lanes; it is very similar to blocking of our coronary artery by deposition of cholesterol.

Clogging of our memory lanes results in blocking of the

flow of our consciousness. Our observation, concentration gets continuously interrupted by imagination and surfacing of many bitter memories of the past. Our discussions, talks and conversations become superficial, sketchy, negative and meaningless. And then, to find pleasure of living, we have to continuously resort to parties, drinks, T.V. and any other news item or conversation, which will give excitement to the nerves.

Such an excited nervous system results in sickness, lack of judgment in business, at times going beyond our limits in business in search of excitement, leading to frustration, gloominess; worry, talking about our own difficulties etc.

This cleaning of memory lanes is possible only by under-standing the eternal law that no unpleasant experience is
possible without some dislike, prejudice, arrogance or attachment within us. Outside situation, which keeps on continuously changing between pleasant and unpleasant experiences is rather helpful to dig out the negativities
from our sub-conscious layers to make them accessible
for cleaning.

Surfacing of any unpleasant incident every now and then or coming across unpleasant people every now and then is the nature's benevolent mechanism – a reminder that something needs to be attended to, to be washed or cleansed. Some inner search becomes necessary to find out the dirty spot and clean the same by apology, forgiving or friendliness (*maitri*) and love.

Forgiving is necessary because an outside person or situation is not responsible for our bad luck. By holding one responsible for our bad luck, we live in a false world, a world of illusion and confusion and imagine a reality which does not exist. This violation of the eternal law attracts punishment, a penalty that follows the law of compound interest.

When we closely examine ourselves, we find that we keep on breaking the natural law by getting up late, working for late hours, we use harsh words where we should be compassionate, sometimes indifference, sometimes anger, certain faults of commission and omission. It is because mother nature continuously forgives us that we are able to carry on with our normal life.

When a disciple asked Jesus whether he should forgive the fault seven times, he says seventy times seven, meaning thereby that an endless forgiving is necessary to achieve reasonable purification of heart.

Maitri (friendliness -love), *Karuna* (compassion), *Mudita* (cheerfulness) and *Upeksha* (equanimity) are four different parameters with which forgiveness is practiced. Starting with equanimity, one ends up with subjective love, a love which does not need reciprocity – objectivity. This is a persisting exercise, producing pleasant results and a charming personality after prolonged, continuous efforts.

Forgiveness gives seriousness, depth, poetry and music to our consciousness. No spiritual practice is possible with revenge lying in any corner in heart.

8

Positive Outlook

*"Our goal should be to look at everything in a positive manner.
The world has both positive and negative forces. Some time
or the other it will convert negative into positive. On our part,
why not remain positive from the start?"*

~~

All spiritual practitioners as they progress on the right path develop keen awareness of everything happening around them but like other unaware people, they do not brood over negative observations which they make.

A mind which broods over negativity develops fault-finding habits, turns into a gutter inspector, a cynic and wherever they go, they find out what is wrong with the people or a situation. Conversation usually turns negative.

A typical case is a party or dinner where there would be innocent talk on what is wrong with politicians, economy, armaments, lifestyle of young generation, recession, labour, global scenario or any other difficulty in business. All this would end in long talks without any benefit.

Brooding over negativity or opening ourselves against too much negativity ultimately weakens the mental defences and one tends to develop a centre-less personality, trying to please others at all times.

What is necessary is to develop awareness about so many positive things happening around us like many saintly persons helping people to develop personal and inter-personal disciplines, hundreds of social workers who are running schools, hospitals, orphanages, old age homes, improvement in condition of roads and telephones, long distance calls, medical facilities, healthy foods, development of parks and gardens at traffic islands, music schools, dance-dramas and host of public utility services which have filled important gaps in many areas of our life in the last ten or twenty years.

No doubt everything new which happens brings its own problems like airplanes which cut short travel-times are subject to delays - sometimes disasters, but mind must constantly meditate on its utility especially while traveling to far away countries. So also increasing comfort provided by railways.

This needs a shift of emphasis on positivity. There is a typical case of house-servants who provide very useful service for 90% of times, making possible a luxurious life-style, parties, dinners, receptions, our afternoon shopping, gossiping but how little a portion of our time with friends is spent on talking good things about them. How much time we get to talk to housemaids with smile and compassion and what interest we take in their problems. Slums where we try to close our nose are the very places, which provide this most needed services without which present lifestyle becomes miserable.

A reversal of awareness is absolutely necessary to progress on the right path. It is not possible to find spirituality, which is synonymous with joy of living, without reversing this awareness; developing awareness in those blank areas, which demand attention.

If at all we start on this path of developing relationship with our servants or drivers by creating awareness in that direction, we should also guard against the attitude of self-righteousness by which we constantly develop a dislike towards those whose awareness is yet to develop in these areas.

A different company, a different life-habit, a different type of reading is required even to get 10% result on this path to positive outlook.

Riots, wars, political unrest, natural calamities like famines, epidemics, earthquakes, cyclone, floods, have been happening to man-kind since time immemorial. During the same time, great spiritual masters like Mahavir, Buddha and Jesus, men of art and literature, great musicians & sculptors, brilliant scientists are also produced. Contributions can be made to relief work in the aftermath of riots; but a meditator should continuously train his mind to be aware of this positivity to raise his consciousness to a higher state.

Garbage to Garden Soil: One enlightened industrialist meditated on impermanence (*anitya*) of garbage and could see that it could change to garden soil under certain conditions. Nature required 40 days of silence and undisturbed work to produce this vital change. He did this. This is classic example of 'awareness of negativity with a positive outlook'.

Surrounded as we are with majority whose awareness about negativity is very powerful, this is difficult task although worth trying, worth achieving, landing one to a higher plane of awareness howsoever insignificant initial gains appear to be.

9

Purification of Desire

"Desire is nothing but dependence on others."

~~

Desires arise in human mind spontaneously. No effort is to be made for generating desires. Of the innumerable desires, which arise and disturb the mind, the most common one seen presently and also written in the same language in which it is spoken is 'to make money'.

People look around for rich people, see their life style consisting of large apartments or houses, cars, servants, industries and many more things and immediately decide that only goal which can be worthwhile is to make money. Even those who are rich want to get richer only to equate with someone higher up on the ladder. Even after reaching, there is no end to it. Higher goals come in sight and the race continues till one gets worn out and tired and then he leaves the race not out of satisfaction but out of frustration, i.e. there is no further energy to run.

Real life examples of those who came out of the race are those who become old and whose children when they join the business do not listen to them, one becomes a victim of serious illness; wife's illness makes him divert a part of the energy in that direction, business loss, quarrels in partners etc. Few are there who voluntarily come out as 'Innings Declared' in the game of Cricket.

Desires other than moneymaking hover around fame, power or position. All these desires are aimed again at one thing, to be happy. Happy with one's own self, with wife, children, health etc. Every desire aims towards happiness. Even when people want to live separate, it is to become happy. Even when they quarrel, they aim at some business or property, which is supposed to make them happy. One of the mothers' desires seen these days is to put their children in first rank in every conceivable field of school activity. Many times mothers run into excessive wear because of this desire.

Although desire to become happy is universal and universal tool to achieve happiness is assumed to be making money, we find that 99% of the people are unhappy. Causes are well known: sickness, disharmony between husband and wife, between father and son, with son-in-law, accidental deaths and a host of other events which make a man unhappy. Pleasures of money are in fact pleasures of palate and pleasures of comfortable houses and holidays.

An extreme view taken by Indian tradition is not to have money and other material happiness. This is as much wrong as the other one aiming to make money.

First essential thing to understand is that money can give us convenience and convenience is needed. But at the same time, if money is earned by using wrong means, unethical means, then it also generates inconvenience by the law of nature. As against goal of 100% convenience, a mixture of convenience and inconvenience is generated, the proportion depending upon fairness and unfairness of means.

In order to reduce inconvenience, misery or unhappiness, which gets produced along with money, first essential task is to purify the desire. This area has not been looked at; very few persons have an idea that to earn money, we can do some correction within. Hatred, prejudice, strong dislike and jealousy are the first parameters to be cleaned up for walking on the road from poverty to richness or wealth. By a strong auto-suggestion followed by meditation; hatred or prejudice is to be replaced with love; and jealousy needs to be replaced with cheerfulness at the prosperity of others.

Rest of the internal purification is a mechanism to reduce inconvenience generated with wealth. One of the most important parameter here is to control greed, which expresses itself in three different forms:
1. A constantly increasing desire to have more money.
2. To pay less than reasonable or get more than reasonable in exchange for work.
3. Not to share our wealth with others – partly as an act of charity and partly to get rid of the excess. A full trust in nature that she will provide reasonably what we want, is

needed to be able to share our fortune with others.
Tendency to hit back, fire our subordinates, punish them severely for minor faults etc., instead of forgiving also needs inner correction. Last of the inner correction is not to be arrogant with money.

Once inner work is accomplished, nature keeps on producing wealth around us to the extent of our own inner cleanliness.

Purification of desire is achieved by meditation.

10

Vision of the World as it is

"The World is the puzzle itself. God has not created this world at all. God is creator of this world is correct by Christian's viewpoint, by Muslims viewpoint, by Indians viewpoint, but not by fact. By fact it is 'only scientific circumstantial evidence.'"

~~

Surrounded by negativities which are in true sense because of negativities within us, we hardly get a true picture even of the locality where we live, leave alone the city, country or the world at large.

When something negative about an important person happ-ens or is publicized, rumoured, talked about, we immediately believe it to be true, connect it with similar negativity of other famous persons, take our mind in the past to project an imaginary picture that past was better than present and ultimately conclude that present times are very bad and are likely to deteriorate further;

a complete imagination. This is the main reason for developing negativity within us.

If some important person is talked about as corrupt, we compare ourselves with him in our sub-conscious mind and conclude that we are not like him. If somebody is arrogant,
we again connect ourselves and say that 'I am not like him'.

When a house servant, driver or a stenographer is changed,
we immediately compare some poor quality of the new member with good quality of the old staff and say that old one was better, although we know fairly well that when that person was in our employment, we never noticed this good quality or if we noticed we never praised it. So is the case with our criticism of practically everyone whom we know. Someone speaks too much, someone speaks too little, someone is not smart, is over-smart, extravagant, miser, systematic, chaotic, selfish, over-ambitious – a long line of negative qualities – criticism goes on and on with only occasional reference to
some good quality of any human being.

This connection, comparison, criticism is a psychological defect accumulated over a period of time when the inner world was not attended to. It was allowed to develop, as it wanted to. Result is wasteland grown inside with weeds over-grown, a few of them even poisonous.

This psychological defect does not allow us to know and experience the bright side of human beings who live in our locality, city and country. The real world is full of brilliant people with great qualities of head and heart, talents of a good administrator, musician, scientist, clerk, peon, worker etc. There are hundreds of brave and courageous people who have the ability to sort out the mess. There are dedicated and obedient people who execute the plans made by intelligent people.

There are good police officers, good businessmen, good teachers, good servants, good relatives and good friends. But we are likely to miss them if we are having a high dose of negativity. World in reality is mixture of good and bad although this division is not strictly right.

As we meditate and keep on meditating regularly with discourses on eternal value systems, slowly we change our vision of the world by becoming aware of positivity in a diversity of fields and later on a small world of high class positive people in various fields develop around us – a world within a world – an oasis in a desert.

Although true, although possible, it requires a prolonged continuous effort. At times there are frustrations; but if we keep on, we succeed. We experience this great eternal truth that changing ourselves changes the world around us and catalyses that portion which is away from us.

Again in the truthfulness of our vision, we will see many like us who are changing within by following different techniques and still achieving the good results. This will be the last and final vision – to see the world in all its positive diversity.

11

Choiceless Awareness

"'Ego' is not an object. It is the belief that 'This is me'. The moment you believe 'I am pure soul' you become devoid of ego."

~~

If anyone observes the natural process of human birth and meditates on it he will come to realize that we have not decided to be born. When we come to the conscious memory, we find that we are born. We did not have choice of our parents, brothers, sisters, our sex, colour of our skin, and host of other factors affecting us as members of this planet, country, township, climate etc. Conscious realisation of this fact is called choiceless awareness.

As we look out to the nature, its cycles of sun-rise and sun-set, phases of moon, timings of tide and ebb, timing and extent of natural calamities like earthquakes, floods, cyclones, heat and cold wave, epidemics and draughts – all these are choiceless awareness.

There is a freedom to study these processes and work out our defence but the occurrence is choiceless.

Inside the human body, to some extent, body weight can be regulated, disease can be avoided or cured, I.Q. can be developed, ability to swim, play, run, play music, faculty of painting etc. can be developed. This process depends on studying and complying with natural laws governing and development of these faculties. Observation asserts the superiority of natural law and limits of our freedom.

There are also natural instincts like hunger, sex, fear, sleep etc., over which human beings have very little control. This leads us face to face with the fact that in 98% of occurrences happening in our life, we are helpless spectators and range in which we can influence our health, our business profits, our family relations, our successes etc. fall within a narrow range of 2% of the total number of happenings. A conscious grasp of this fact is sufficient enough to pulverize our inflated ego-system that "I can do this and I can undo this".

However man is not born to be a helpless victim of a natural process of birth nor only a clog in the natural wheel of birth and death. He has been able to design rapid transport systems like cars, railways, aeroplanes, fast telecommunications, new type of comfortable houses, foods, drainage, water systems, internet, smart phones etc. by studying the natural laws and devising systems to conform to this law. All this has been possible by studying the external working of nature.

There are many areas in our emotional nature which give rise to health or disease, happiness or misery, wealth or poverty, our emotions of love and hate, likes and dislikes which can only be controlled by studying laws governing our internal systems, becoming aware of them and finally complying with them to devise an internal system of transport and telecommunication, new type of 'Internal House'– beautiful and elegant; this requires a conscious effort under the guidance of a teacher.

First reality we comprehend while studying our internal nature is that what we say 'I went to the office', 'I did this work' , 'I directed the course of a particular event' – all this 'I' is only illusory and some unknown natural force compels us to do all these things, sometimes resulting in failure, sometimes success and more frequently a mixed result.

It is possible for human being to devise a system where he can increase his happiness and reduce misery, increase the extent of his success and decrease failure. The teacher is required at this stage to guide the student how to achieve this and it is not possible for anybody to work out the plan himself.

Observation of natural processes definitely point out to the presence of divine work force within and without. Process of understanding its existence is the awakened state of our consciousness. It makes us humble and energetic. Directing our attention to the dissolution of 'I', brings us an awareness of 'Sleeping State' during which we have developed eight-fold impurities – hatred, attachment, greed, strong sense of possession, revenge, jealousy, anger-arrogance and expectation.

Conscious effort for correction converts these to positive emotions of love, charity, cheerfulness, compassion, humility and acceptance. This change converts misery into happiness, poverty to wealth, disease to health etc.

This process of understanding illusory 'I', becoming aware or experiencing its working and ultimately subliming this 'I' alternatively known as ego-principle is called ego-less behaviour, self-realisation, God-realisation, divine knowledge, *Brahma-Vidya* or awakening of the soul etc. Practice involves converting instinct into well-trained and controlled reflexes thereby developing divine insight into the workings of natural processes, external and internal.

12

Illusion

"The world is the puzzle itself, but it is always in principle."

∼∼

The most important illusion in a natural process is lateral inversion – familiarly when right side appears left and left appears right, causing the phenomenon of day and night which appears due to sunrise and sunset but in reality there is no sunrise and sunset, rotation of earth on its own axis, causes illusion.

Corresponding example in human system is coming in contact with unpleasant experiences like meeting a bad land-lord, a business partner who does not remain faithful, quarrel between the partners, unpleasant experiences to a woman in early years of marriage etc. In all these, nature creates an illusion, by making somebody responsible for our bad luck, but in reality the causes are within. Illusion is so strong that man goes on fighting, hating and ultimately hurting himself, wasting energy or spoiling health. This illusion can be cured only by spiritual masters.

Similarly government and politicians we get are our own projection of hatred and greed, but illusion shows that politicians are self-seeking and fighting between themselves. Whenever there is labour trouble, we try to identify the causes outside but the real reason is our excessive greed and unfair treatment.

A process of fasting or semi-fasting in essence is a process of cleaning to obtain pure energy at the end of experiment, but illusion shows initial weakness and cleaning process can be experienced much later.

When we work either at office or at home, more than 75% of work is unnecessary, it is more noise than work; silence prepares the system for a more efficient work but apparently it seems that nothing is happening in silence.

Indian life is full of impulsive explosions. There are explosions on trivial things. We make too much noise showing too much turbulence in our minds. We speak much more than necessary. While reading a newspaper we put too much information in our minds, which is unwanted and useless. Result is that life outside is full of noise-undisciplined traffic, noisy festivals, and crackers during marriages and *Diwali* (very sharp and harsh sounds) – all this noise and turbulence is because of noise within us but through illusion, the cause appears outside. As a result of meditation, when inside becomes quiet and peaceful – in that condition, although noise remains outside, we are not disturbed. We interpret this noise as originating from disturbed minds and have compassion and goodwill towards them. This is called coming out of illusion.

Through illusion, emotion appears as feeling and love whereas it only projects our weak mind on our relations and makes them weak. In fact, we should be in motion and not in emotion. Emotion is something like derailment where all motion stops and only confusion and anxiety remain.

Another famous example of illusion is the difference between religion and spirituality. Whereas spiritual practice brings us in harmony with eternal laws of nature and most of it is unseen consisting of subtle psychological changes, religion consists of certain do's and don'ts. Almost all rituals can be seen. Majority of the people mistake religion as spiritual practice and believe this to be the aim and end of all things they are seeking as spiritual practice.

One of the marvellous illusions created by nature is regarding non-violence. Physical non-violence can be easily understood and practiced. It seems this is the end, whereas real violence is inside- hatred, greed, arrogance, deceit, jealousy etc. Inward effort is necessary to become non-violent. This being not so apparent is not being practiced. So also for earning money, inward effort, non-violence, automatically produces wealth. Through illusion man makes an outward effort.

Under the illusion, law of growth & decay is not seen and craving develops for relations to remain permanent. Life in the village, which is highly inactive, appears peaceful. Body, which is seen, appears to be a real body. Clinging develops for preserving the body. Unseen body, which does not die but gets developed during repeated cycles of birth & death, is not grasped. It is this body, which requires major attention. Fearful imaginations appear as thoughts, where anxiety gets justified.

Lessons from experiences of pleasure and pain cannot be understood. Gradual nature of natural changes cannot be perceived and wrong decisions are made. Possessions appear as safety. Attachment appears as love. Calculations are made out of fear and right practices are abandoned. Whole life becomes jumble of illusions, confusions get worst, confounded chaos and contradiction dominate speech & action. Life without love, relationship without reality and death without grace is the final outcome of human beings under the grip of illusion. Hindu scriptures call this *maya*.

Impermanace

*"One who knows and says temporary as temporary,
should be permanent."*

~~

There has been a widespread pessimism developed in Hindu society citing *anitya* as the principle philosophy of life thereby meaning that everything is impermanent; and there is no permanent basis for relationship. All that we see is subject to destruction and death. People are not encouraged to put up any creative effort, or hard work, because all material wealth is considered impermanent and accompanied with misery.

It seems this is a wrong interpretation of impermanence. Truly it means that any difficulty we are in – emotional, physical or in business situations, is of temporary nature and is subject to change. As many of natural processes are cyclic in nature, subject to law of growth and decay, concept of impermanence has to be mathematically treated by determining limits, direction of change and treat this change itself as a function of time.

Viewed in this light, all changes are subject to a limiting value and involve change in levels of energy. During meditation, energy levels change their direction and go upwards, whereas in normal life, they go downwards.

When we consciously train our mind not to become victim of hatred or prejudice, resist tendency towards jealousy and greed, mental energy level representing memory and concentration increases and under condition of free flow it decreases.

A system in which energy levels are under the control of nature i.e. life principle or soul is in dormant state, changes in energy level or corresponding states of purity follow a wave like motion going above and below the reference line with a finite amplitude, unique for each system. This is a development path. Under conditions of awakened state of life principle, over and above the normal wave motion, there is a steady rise in energy level. In the development path energy goes up and down, whereas in the awakened state, it only goes up in a wave like motion.

Another interpretation of *anitya* is that entry and exit of impurities or changes in energy levels have discontinuous continuity. They do not increase or decrease in one step. There are quantum jumps followed by periods of rest and after a time continuation of this activity. Alternatively, between periods of two unhappiness, there is a period of happiness. Thus, gradual nature of changes in natural cycle and system of alternative pleasure and pain make the development path more bearable. This is the greatest blessings of nature.

Another meaning of impermanence is that either happiness or misery, pleasure or pain, favourable and unfavourable situations are not static but dynamic in nature. Pleasant experience changes into an unpleasant one at the end of its time-cycle and unpleasant changes into pleasant in a similar way. This gives a theoretical basis for meditation not to generate craving or aversion towards people or situations because it is impossible for any pleasant event to remain for a long time and vice versa. Even if we do not crave, a pleasant situation will come at the end of an unpleasant one and the unpleasant cannot be avoided however strongly we wish. By remaining neutral during an unpleasant event, or trying to do so, nature produces strong life currents to bear a painful situation with progressive ease and this strength becomes useful in good times to resist craving.

Development of this strength is the aim and end of spiritual process of a natural cycle. Meditation only accelerates this evolutionary process. Understanding the concept of impermanence leads us to freedom from craving and aversion and establishes human beings into a spiritual centre line, which is the essence of existence alternatively known as love.

One more interpretation of *anitya* is that everything in nature outside and inside is in a constant state of flux, supported by vibratory motion throughout eternity. This state of flux generates infinite number of patterns of behaviour, moods, sensations, relationships, experiences etc., each one distinctly different from its predecessor. *Anitya* means non-repeating wave like nature of pleasant and unpleasant experiences. Entire concept is meant to come out of attachment – one of the principal causes of misery and suffering.

The most important interpretation of *anitya* is that all pleasant and unpleasant experiences are caused by nature and have a creative purpose. Only through illusion one finds either situations or people responsible for our difficulty. By remaining neutral and observing sensations during meditation and the mixed patterns of experiences in actual life, it is possible to understand the purpose of nature, i.e. to take us out of our cravings and aversions bringing us to a centreline of our existence where infinite energy, joy, peace and happiness are experienced.

14

Observation & Pattern Making

"Charging (of Karma) is done in human form, discharge is done through natural regulatory system."

~~

It is difficult these days for people – who are in the imaginations of future: worrying, afraid of so many odd happenings, living with tense nerves or in the memories of the past either with stored hurts or glorifying the past; to develop faculty of observation. Their mind is always pre-occupied with unpleasant past or uncertain future.

However, as we empty our minds with regular meditation and are able to keep our attention in the present, we are able to make a number of observations about ourselves as well as about the family members in whose midst we live or the staff members with whom we work.

Observations about us show that our moods keep on changing from hour to hour. Morning we may be in a neutral mood, but reading a newspaper may make us tensed or depressed. In the office, the mood is mixed one of rejoicing, anxious or tense, firing, quarrelling, fighting, getting annoyed with repeated frustrations, getting fed-up with insubordinate staff or over-controlling boss, flattering or getting fatigued, tired, worn out, inconsistent working, planning, getting excited over success and develop craving for more success etc.

If we observe ourselves for one full year and generalize, we might have prospered in our business or sometimes lost; either a good year or a year of family sickness, partitions between brothers etc., some unforeseen tragedies like riots, earthquakes etc. Same pattern repeats if we take a period of ten years except that either some new children may have been born or some elderly family members might have passed away. In inter-personal relations, new people and situations enter into our lives for a period of one to five years then changing, resulting in bitterness, fights, greed, hostility, triumphs and tragedies – less frequently sweet memories of associations or groups of friends from whom we have parted.

All these patterns fall into two broad groups:
1. Our fortunes, good luck and bad luck alternate one after the other and are within certain definite limits.
2. Super-imposed upon this is a general graph in which certain people lose their prosperity slowly and step by step and reverse is the case with other set of people who prosper slowly step by step.

A mixed pattern of failures, frustrations, tragedies, successes, fulfilments, and triumphs weave this fabric of increasing or decreasing prosperity. Most people feel wiser and developed every ten years including those who lose and those who gain.

After the age of 40, if people analyse themselves, they will find themselves in a distinct pattern of either hostility to the people, government, society or anxiety neurosis – usually a mixture of these two with one of them more predominant.

Husband and wife form complimentary pattern; what is lacking in the husband can be found in excess in the wife and vice-versa. Two together if wisely adjusted and under-stood become a most fruitful combination. Such is also the divine purpose.

Two brothers usually form a heterogeneous supplementary pair. Infighting between brothers or their children is a very common pattern – nature's way of separation and re-integration in different groups.

There are well-defined patterns of restless and ambitious people; they wear out very fast. There are also excessively greedy people who destroy the best part of their personality in going for more and more money.

When we observe human beings in large groups, we find that majority falls in happy-go-lucky normal men and women. Percentage of out-of-ordinary people keeps on decreasing as the extra-ordinariness increases and becomes microscopic at the top-most level.

This is the development path – misery and impermanent nature of misery; happiness and impermanent nature of happiness. Misery or unhappiness generates an inward pressure to look within and correct the weak links in our thinking pattern &

happiness does the reverse.

Getting fed up with this alternating cycle of happiness and misery, many people take to a spiritual path; either rituals, worship, meditations, study of scriptures, company of holy people etc. This gives them solace, wisdom and strength to go through boredom and sufferings of old age. This is the purpose of our birth.

15

Divine Patterns

"What is the purpose of speed breakers on road? It is for your safety. Likewise obstacles or difficulties are like speed breakers otherwise one may have to face an accident because of high speeds. Obstacles help to live with normality."

∼∼

Nature produces infinite number of life patterns in infinite number of men and women of all races and religious groups. There are people having greed, arrogance, jealousy, attachment, generosity, humility, love, cheerfulness and a mixture of all these. Infinite number of positive and negative emotions and their mixture in an individual in various proportions produce a multitude of patterns in such a way that no two human beings will have identical behaviour. Behaviour may approach near similarity for some time but never an identical.

Other important characteristic of these patterns is that they keep on changing noticeably over a period of time. A man who was careless and extravagant becomes an alert controlled person; misers realize the futility of greed and get frustrated, arrogant get entangled into bitter quarrels and fights and at times becomes humble and vice-versa. People who are born rich become poor at the end of life and those who were poor become wealthy as life proceeds.

One of the most frequented patterns, husband-wife relation, changes from attachment to misery and hostility as the marriage proceeds. In number of cases hostility remains till the end, producing a mixture of hostility and attachment.

Usual patterns of people in pairs will be – optimistic- pessimistic, supportive-fault finding, enterprising-timid, clever -dull, wise-foolish, trusting-doubting, active-lazy, impulsive -cool, open hearted-close minded. You can expand this list as much as you like and it will be helpful to know that every pattern produces its opposite.

Business or industrial organization will invariably consist of ambitious boss or group leader. Growth of a business depends upon the ego level of the boss. Lower the ego level, larger the expansion and vice-versa. Ambition is an absolute must for development, sustenance and expansion of business. Ambition is a positive parameter. Depending upon the ego level each organization will have three groups – one pro-management, second neutral or slightly passive and third anti-management. This anti-management or conflict group is a measure of the ego level or arrogance of the boss. This is a universal pattern in every organization. Making an organization successful and lowering of the ego level is like two sides of the same coin.

One common pattern that we find in majority of lives is that as one crosses the age of 50, inspite of all plans for happiness, one feels empty and miserable. He or she realizes that children do not listen to him anymore, there is bitterness in family relations, there is expensive sickness, at times premature loss of loved ones, one thing or the other. It is theoretically impossible for any life to be without failure, frustration or a minor/major tragedy by the time one reaches the age of 50 or 60. These experiences draw a man inward.

For interpreting these patterns, we have to work with one hypothesis, i.e. whatever happens in nature is with a purpose and the only purpose for which nature works is to change our attachment to love, our misery to happiness. Although experience of failure or frustration is common to most people, interpretation of this experience being caused by attachment is not so evident. It requires a third party – possibly a teacher who has been partly successful in overcoming his own attachment.

As a result of failure or frustration – people take to worship, study of scriptures, repetitive chanting of holy name, prayer, meditation, pilgrimage, singing of hymns, listening to sermons; one or more of this in search of life beyond sorrow, misery, failures and frustrations. Whether they know it or not, they are sowing the seed of a better new life after death.

As the life span nears completion, everyone experiences a supernatural force beyond their control, which shapes events in life that has passed and they learn to surrender; their ego gets pulverized. Apparently, it looks as if the people are getting helpless. It is true also for some time; but in reality, they are coming in line with the divine power. Although process of pulverization of ego does not get completed in one life cycle, step-by-step it brings an individual in tune with outside divinity. This in turn arouses inner divinity and at the end the difference between outside and inside vanishes and the whole consciousness becomes infinite, with love, peace, joy and harmony.

This is the ultimate pattern – a pattern beyond all other patterns, a pattern mysteriously divine, all knowing, all pervading.

16

Rhythm & Balance

"Material pleasures which are above normal lead to unhappiness or sorrows."

~~

With continued attention to the external world, human beings have lost faculty of turning attention to internal world whose space and variety is as big as the world outside. Loss of this faculty has generated a heavy tilt in consciousness with decreasing control of instinctive impulse of sex, hunger, fear, anxiety etc.

This loss of balance results in loss of rhythm and harmony progressively going down to disorder, turmoil and chaos. There will be infinite number of waves arising in human mind under the effect of turmoil and chaos, consequently mind will be weak and receptive to outside infection. This is the origin of all diseases of all types.

A chaotic system is one where the mind becomes dull and inactive. This system is open to all addictions. It also has in-built violence towards other people and women. This system is not very conducive to yogic practices. It prefers to worship God for some reward or protection.

A mind in turmoil represents varying degrees of hyper-active systems. Most people in this range represent over-developed or under-developed ego system. Majority of this group are married and take to some economic activity. They show a high degree of attachment to money, material wealth, wife and children. As the life span progresses, they experience considerable frustration because of attachment and take to spiritual practices – either rituals or yoga or scriptural readings.

People having minor loss of balance giving rise to a disorder in the system are lesser in the society. Many of them remain unmarried throughout their life. They have less of attachment or have strong dislikes. They have anger, arrogance as main defilements and go for name and fame. All *sanyasis*, priests, ambitious people, political leaders, fall into this group. Many of them take to advanced spiritual practices under the guidance of an enlightened soul.

Advancement in the development path may take both the directions like pendulum, disorder to turmoil and disorder to order eventually it takes one upwards at the end of the development cycle.

Path from chaos to turbulence although long enough, is

relatively easy, compared to path from disorder to harmony. Here there are too many temptations – wealth, fame, luxurious living, flattery by disciples etc. At times, beautiful women cross the path of advanced yogis; at other times anger accompanied with arrogance. These forces are like snow-storm while going up a high mountain. Sometimes people get lost in these temptations to come up again after a very long period.

As against turmoil and chaos, rhythm and cosmos is ultimate human destiny and also birthright of every human being at one time or other. With temptations and pitfalls, nature takes every human being to perfection – a state without disease and decay; state of rhythmic dynamism; a very powerful state, silent and serene, at times appearing static.

A movement from chaos to cosmos, turmoil to rhythm is called real effort; all other efforts are only turbulent movement producing very little reward for a very large work; most of the work being lost in eddy currents.

A human system approaching a progressively purer rhythm, experiences ever increasing vastness of consciousness touching astral regions. Turbulent people experience peace in his presence. Voice becomes ringing and sweet; face serene and radiating. Memory becomes sharp, voluminous, varied and prolonged. People with rhythmic minds are able to read the thoughts of others and also develop perception of distant events.

17

Uniqueness of Experience

"Scriptures are in the nature of a 'helping problem'. They give direction. They act as a thermometer. A thermometer cannot replace a dose of medicine! Can it?"

~~

Nature has designed us so that although all of us have two eyes, one nose etc. each one of us is unique with respect to rest of us. Two faces may be very similar but there is something unique which helps identification. Also, the voice, style of walking, working, dealing etc. No two signatures would be alike, not even one in a billion.

This is the method which nature uses in producing variety and uniqueness in human beings although general structure is same. There is a racial variety all over the world, with uniqueness in each one of them.

There are nine prominent negativities with their nine positive counter-parts. They are hatred-compassion, attachment-love, greed-generosity, hypocrisy-straight-forwardness, strong sense of possession-charity, revenge-forgiveness, jealousy-cheerful at the prosperity of others, arrogance-humility and expectation-acceptance. These positive and negative qualities are of varying degrees and their interaction produces infinite number of experience patterns, each one unique by itself.

Partial, slight, predominant or major change of negativity to positivity generates a spiritual experience from modest to tangible and vice-versa. Because of infinite variety of experiences, no two individuals even husband and wife will have similar experience or a behavioural pattern on a development path. Although living under one roof, eating together or listening to same discourse, experience will be different and distinctly unique giving rise to a variegated pattern.

Because all spiritual experiences unite at the highest level to form a similar and uniquely distinct pattern, Human beings crave for similar behaviour pattern in inter-personal relationship. Husband expects wife to behave similar to him and vice-versa. So also for two brothers, partners, friends, neighbours etc. This is the power of illusion. This causes stress, quarrels and bitterness in relations. Although aim is to love, there is hatred, aim is to unite yet there is separation, aim is to co-operate yet there is quarrel – blind man leading another blind.

Understanding the uniqueness of experience and corresponding behavioural patterns, progressively resolves fights, quarrels, dislikes into understanding and appreciation of uniqueness of life pattern.

It is a common experience that every human being will have brothers and sisters, some of them he will like and others not so much. In case of neighbours, business partners, colleagues, workers, staff, government officers etc. there will be some who will be favourable and others unfavourable. After learning uniqueness of behaviour pattern, we turn to a meaningful interpretation of these patterns, each unique in itself. A careful study will show that all favourable experiences correspond with our positivity and unfavourable ones correspond with our negativity.

All our negativities are projected in the outside world in the form of enemies, opponents, cynics and trouble-makers. If we make an attempt to fight them, they will have a tighter grip on our necks. They will pulverize our ego and make us helpless. If we make a mental effort to love them, then they start disappearing one by one. This is the secret of creative spiritual path.

Knowing uniqueness of each spiritual experience helps us to develop better and sweeter human relations. Our expectations will change to acceptance – the most powerful basis of spiritual practice. There will be no more prejudices, annoyances, irritations, tensions etc. There will be joy and peace in watching the unity in diversity of all experiences.

As one progresses on this path, heterogeneous systems will change to homogeneous complimentary ones. More of such people working together will weave a pattern of different colours each blending into the other to make a meaningful picture. All such meaningful pictures, negative and positive, have a purpose in nature's development path, negative-painful and positive-joyful.

All experiences when they are devoid of any negativity, will be uniquely similar and serene and to have similarity of this positive experience is eternal human desire, its ultimate destination.

Spirituality & Diet

"One should eat in presence of mind. Eating absent-minded is a major cause of disease."

~~

The aim of any spiritual practice is to exchange our negativities of hatred, greed and violence etc. into positive emotions of love, compassion, charity, humility etc.

Change of this emotional balance requires energy – a very subtle psychological energy which generates auto-suggestion either to pacify or completely remove these negativities.

During the initial years, aspirant generates this energy, either gross or subtle from the food he eats. If a seeker eats pungent, salty and stale food, he generates type of gross energy like any animal, not suitable for any subtle intellectual or spiritual work. At times man eating this type of food is dull, lazy and insensitive to emotional disturbances.

A man whose diet consists of simple food with moderate intake of salt generates a mixed, brain energy, a mixture of positive and negative emotions. This is useful for day to day work but does not help decreasing negativity.

A diet consisting of moderate amount of fruit, fruit-juices, non-spicy cereals, leafy vegetables (raw or boiled) and little of milk, butter, ghee generates spiritual energy necessary for the change. One of the important energy required during the initial stage is control of sex, an outgoing energy. This is done by meditating on the temporariness of this pleasure. Also, a strong auto-suggestion to restrict this urge and see what experience one gets. Energy required for this exercise comes from this type of food. Severity of the impulse and restlessness gets reduced making the control easier.

As the energy conserved due to sex goes upward, the body becomes supple and energetic, senses of touch, hearing, observation become sharp and distinct, memory gets enlarged retaining varied incidents over a long duration, voice becomes sweet and melodious, mind becomes alert and sharp-witted to make patterns of observation. Anger can be controlled and arrogance can be understood in detail.

With food more-or-less remaining same, the boiled or raw with minimum frying, the next psychological battle for higher experience is against praise and flattery. Aspirant is able to rise above the average people and do some outstanding things. People praise him or flatter him. Fasting is added as supplement to pure food to burn these impurities faster. Clear perception of aim and strong suggestion to come out from negativities help to sustain this diet.

Control or awareness of breath alternatively known as *pranayama* or *pranapan* is necessary to make energy release evenly in all parts of the body as every cell of the body comes to life.

As the seeker further progresses on the path, a very subtle food of meditation is necessary to come out of craving and aversion; since every situation we come across in the world outside whether pleasant or unpleasant is a projection of our own mind and the only way to be happy and peaceful is to change our psychological make-up. This higher state of consciousness requires pure food, solitude, meditation and discourse with the teacher. All these are foods – gross to subtle, subtler, subtlest.

Breathing is a very subtle food. It purifies the blood and carries life giving energy to nerves, blood, bones and tissues. Breathing-in of oxygen is by pulsating motion of lungs. Yogis are able to generate pulsating sensations throughout the body and thus breathe directly too.

Regular meditation develops an ability to digest the subtlest of all energy food and solar energy, directly from Sun, the source of all energy. Whatever food we eat, ultimately it is the solar energy gathered by the plant and passed on to the grain or fruit which supports life at various levels of our consciousness. Pure and direct solar energy is the highest and subtlest energy food.

19

Independence & Interdependence

"Religion begins with obliging nature. Mutual obligation is privilege of human life."

~~

What we usually understand by independence is that there is nobody to tell us if we come home late or get up late or do not go to office in time or spend money as we like, or eat more spicy food at later hours etc. So also for smoking, drinking and other habits. What goes in the name of independence are uncontrolled life habits.

Real independence starts with an effort to bring life habits under control. Go to bed in time, get up early, be punctual in work, moderate in eating and talking, all these are preliminary exercises at self-control which progressively leads to independence.

Indulgence or uncontrolled life habits leads to loss of energy and ultimately disease. Controlled life habits lead to health and vigour. Independence here is from disease.

A human being whose happiness depends primarily on comforts and secondarily on the condition that everything should go as he wishes, is creating a conditioned personality; if he develops dislikes, prejudices etc., he creates a strongly conditioned personality open to weakness and disease.

True independence consists of equanimity of mental state under reasonably uncomfortable situations and step towards such progressive independence is an understanding and practice of law of acceptance, which requires graceful acceptance of all situations as the only path to independence. Secondary qualities of such independence will be lack of hatred, jealousy, greed and arrogance. Behaviour, which is not hostile to an out-side situation makes us move towards independence.

Constitutional right of freedom of speech in effect leads to uncontrolled talking, back-biting, malicious and negative talk, gossiping etc. Real freedom comes from controlled speech accompanied by humility, humour and creativity.

A human being who becomes victim of lust or greed progressively loses his independence in favour of cravings, name and fame. Very often sleeplessness is the result of such cravings.

Control, independence and non-indulgence is the ultimate human destiny. Other divergence is only a temporary variation.

In the entire working of nature, there is no such thing as independence. Independence as mentioned above leads to a conscious awareness of our inter-dependence on wife, children, friends, partners, neighbours etc. No life can exist alone. We are also constantly dependent on the balance of nature's five elements, which support our existence. Building a relationship with nature with thanks-giving is a sure path towards independence. Path although hidden, does really exist. Meditation opens the gate of this secret path.

A human being, who subjectively finds the method of loving friends and opponents, becomes consciously interdependent and truly independent. This is the aim of all inter-personal relationships.

Economic Activity & Competition

"Dishonesty is the Best Foolishness."

~~

In today's world, economic activity is mainly concerned with production of goods, improving upon its speed and quality aimed only at the profitability. In the process of increasing profit, if the process harms the nature or human being, or results in slums and problems of addiction etc., it is considered to be of marginal importance. Aim of an economic activity is said to give maximum employment with maximum possible
pay packet.

As the economic activity does not consciously touch upon the negativity of human beings and its correction even as its secondary aim, only a slow natural process takes place in the development of man-kind.

An economic system whereby human being is being continuously fed with ever increasing desire to have more and more material comforts, results in weakening of human mind with resultant fear, excited nervous system followed by multiplication of medicines and diseases; sex, violence and tranquillizers become part of every-day life. To add fuel to the fire, competition wants to throw out an old business into extinction, consequently sense of insecurity and helplessness is added to the above referred evils.

Nature does not act like an absentee landlord. It tries to restore its balance and this creates a new horizon of spiritual approach to economic activity. In this system, conventional method of economic activity – namely production of goods and services is given secondary importance. Primary concern is to purge the men of their hatred, greed, jealousy and violence. A systematic training programme in theory and practice needs to be undertaken with clear target in sight.

Once this takes off in a reasonable way, mind generates sufficient energy from within to keep itself cheerful and excess material needs become unnecessary to keep oneself happy. This includes luxurious expense on house, cars, dinners, parties, drinks, TV, late night parties, ornaments, farm houses, beach resorts, private air-crafts, foreign trips, excessive marriage expenses etc. As the healthy mind does not become sick frequently, excessive medical bills are also cut-off. As a result, decrease in all these expenses requires less material goods to keep one happy. Modern necessities like car, telephone, house, refrigerator, TV, traveling etc. are not eliminated but excessive requirement gets curbed. This is the first step in spiritual approach to economic activity.

Once the consciousness becomes pure or starts off from impurity to purity, nature provides whatever we want; house, money, factory, technology, job or any reasonable necessity to give effect to the nature's task of producing goods and services to make life happy and enjoyable. Very little effort is required to sustain this production. It almost happens automatically. It is stress free. It does not depend on flattering or pleasing anybody. It does not depend on government policy. Only condition which sustains this activity is the purity of consciousness, which in turn invokes the divine powers of nature to sustain it. It grows and develops by itself. It is self-sufficient, self-sustaining, and independent of whims and fancies of government and vagaries of nature.

In this system, there is no water tight compartment between spiritual activity and economic activity. Spiritual activity means cleaning ourselves of negativities and this supports economic activity. Material prosperity in turn provides necessary infrastructure for internal cleanliness. In the end, it becomes a chain reaction. Such is the grand design of nature – unfoldment of divinity within men and women.

When economic activity is based on spiritual principles and in turn helps in unfolding divinity, competition becomes meaningless. There is enough for supporting spiritual economic activity of entire mankind by producing adequate goods and services. Nobody needs to be wiped out. Only organizations who are not able to continue are those who are not on the right path. Competition becomes superfluous as there is room for everybody to exist. Instead of competition, there is a desire for subjective excellence. Such an economic activity radiates peace and joy, is truly inter-dependent in the sense that it supports nature and in turn nature supports its children, an eco-friendly economic activity.

21

Human Energy

"Egoism consumes all your powers; in addition causes you harm."

~~

All human beings have some mechanical energy enabling them to do manual work of farm or factory labour or domestic servant. When subtle mental energy resulting in increased concentration is added to the existing mechanical energy, man becomes a craftsman like fitter, welder, mechanic, carpenter, black-smith etc. When call on mechanical energy is less and brain-energy requirement is more, man becomes a clerk, telephone operator, draughtsman, cashier, accountant etc. When mechanical energy is supplemented with brain energy, value addition starts taking place. Compared to a helper, a clerk is valued more.

As the brain energy becomes finer, it produces a doctor, an engineer, a lawyer, a solicitor, a businessman etc. This requires concentration and understanding. Understanding represents finer form of energy and value addition is more. In between concentration and understanding comes memory, planning, imagination, direction etc.

There is also finer energy, which is required to control the natural instinct of craving and aversion and calm down the waves in the mind. This controlling energy is the subtlest and fetches the highest value, partly material and mainly spiritual. Manifestation of this energy is prolific memory, sharp observation, creating patterns of observation and interpreting them. This energy produces superb powers of healing and helping people to come out of negativities. The energy is having infinite value addition making it invaluable – beyond any conceivable value.

This final energy, where whatever one wants becomes easily available and whose gross manifestation is skill, memory, calculation, observation, imagination, understanding, etc. remain nearly frozen in the case of a slum dweller, house servant or a helper. It becomes involved, sunk or dormant and is also called hidden, latent, sleeping, coiled etc. When energy gets involved and dormant, people look dull, stone-like but still have the same potential in some future births to be perfect human beings.

When the inner energy becomes nearly sunk or involved, the gravitational pull of nature, which attract everything downward super-imposes this involved, apparently extinct ocean of energy with negative emotions like hatred, attachment, greed, arrogance etc. along with their bye-products like fear, nervousness, prejudices, flattery, obstinacy etc. Over a period of time, this super-imposition becomes so natural that we feel this is our true nature. When this sunk energy is about to open up, it produces a hyper-active pattern where characteristics of pure energy, which are love, compassion, forgiveness etc. get mixed up with negative manifestations like hatred, greed etc., and produce dichotomy of a mixed pattern.

Misery is the result of this mixed pattern and meditation is the solution. Any type of meditation, whichever type is suitable to one's temperament, will unwind the coiled-up inner energy level. By law of nature, during this unwinding, he will meet a saintly person who will also inject divine wisdom in him according to his capacity. Energy and wisdom together will then manifest in the form of love, compassion, fearlessness, equanimity, forgiveness, charity, insight, understanding of external truths, receptivity to subtle vibrations etc.

Uncoiling and unwinding of this potential energy is the biggest effort of the soul, producing mysteriously beautiful men and women. This is the ultimate human destiny.

22

Husband Wife Relationship

"Your Spouse is your counter-weight."

~~

Of all the dualities produced by nature, duality of man and woman when paired as husband and wife is unique and spiritually most useful. Very few can grasp the constructive purpose of this duality paired during the life-span of one of the party.

Complimentary nature of marriage brings together two persons having same level of nervous (neural) strength but contrasting emotions of fear, greed, anger, jealousy, attachment etc. If husband has well developed anger, wife would be calm but will have a strong sense of attachment. If husband is generous, wife would have narrow mind. A talented woman would have dull husband, adventurous man may have timid wife, and vice-versa.

Degree of reversal will depend on development of ego system, usually a male partner will have over-developed ego system, which will be manifested by dominant behaviour. More dominant the behaviour of one of the partner, more distinct will be the reversal. Reversal has a purpose. An egoistic person will come into some difficulty or other because of the ego. Either wealth will be a problem or health or at times both. When this happens, other party with less developed ego system will be able to protect him or her.

In times to come, partner with under-developed ego system also starts developing creative instinct of talent, courage, skill etc. Other partner whose ego has gone down will start protecting him or her. This type of complimentary exchange takes place when any one of them starts working for development of higher self.

At times, husband and wife who are strongly attached to each other start fighting and quarrelling. Here the nature wants to develop each one of them by reducing excessive attachment and hence the fight. In the development path consisting of numerous cycles of birth and death, in one particular cycle husband-wife relationship develops both of them by generating negativity in each partner to come out of excessive attachment. They may even think of parting the companionship. Sometimes, they do not like to see each other but ultimately as death comes nearer, each person is able to see the good points on the other side and generate a loving relationship. This is the purpose of marriage – starting with attachment, it ends with love and interdependence.

Divorce is a bankruptcy of wisdom on the part of both the partners. Any remarriage after divorce to a great extent results in increased misery. Whenever, there are difficulties in marriage due to one partner going in a wrong direction, it is the opportunity for spiritual development of the other partner. By becoming compassionate on the erring partner, we increase our strength and the vibrations of love are generated so that it has corrective effect on the other partner. Broken relationship leads to broken life.

Husband-wife relationship when understood properly under the guidance of a good teacher, experimented with positive behavioural responses becomes the powerful tool for achieving divine union – divine in the sense that either attachment or strong dislike, fear, jealousy, frustration, obstinacy, etc. in each partner stand considerably reduced after about 30 years of married life. Both recognize that every crisis in health, wealth or family relation is supported by either partner, and realise that indeed they are made for each-other. Whoever remains behind might experience that fulfilment.

Happy and enlightened husband-wife relation acts as an umbrella to sons, daughters and their spouses. Younger people have energy, but they want wisdom. Such wisdom in parents and grand-parents comes as a result of happy marriage. Initially wife wants protection and later on husband wants protection in times of crisis. Ultimately both protect each other and thus fulfil nature's role of protecting every human being so that they realize their own divinity.

Successful husband-wife relationship is also the basis of successful economic activity.

23

Mystery of Death

"For those who do not believe in rebirth, there should be no word like 'fate'. From where the words, lucky-unlucky, taqdir-tadbir, come? They suggest previous connection. Language of different faiths is complete but the belief is incomplete."

~~

The way nature works, every activity of nature is kept hidden. Development of child in mother's womb is not known either to the child or to the mother. Various growth processes that take place in the body: sleep, digestion, old age, disease – every one of these activities takes place in silence and is covered up, so that no one knows when these processes actually take place; what are the underlying causes etc.

Sometimes wise people get a semblance of the causes and by controlling causes, they also control or manipulate the effects. Many times, controlling energy which is weak, is stepped up by a meditative process. In fact, there is no mystery except that hidden causes have been found out and by setting certain positive causes into operation, certain unusual positive effects are obtained. Similarly, by stopping the onset of negative effects like fear, fanaticism, frustration etc., events like disease are avoided or cured. All such happenings are termed mysterious.

The mystery of death falls within the scope of such hidden operations of nature where the cause is carefully hidden and gross effect is only visible. However, wise people who have solved this riddle, have professed to others who are not capable of solving the process of death that it is similar to discarding an old car which requires too much maintenance in favour of a new car, nothing more than that. Model of the new car may be different, at times inferior to the old model, at times superior, but every time it is new, old has become too old, not usable, sometimes too dirty and monotonous and new is fresh and invigorating.

Fear of death is because we feel that we are finished – become extinct. No one wants to be extinct. Every human being instinctively feels that he should remain permanent and it is really so. However, mystery of death has to be examined along with mystery of birth. Breath currents put into operation by nature at the time of birth are so powerful that the soul is just put to sleep and pseudo soul (Dada Bhagawan calls it *pratishthit atma*) is created which believes that this body is myself, or I am the body. These pulsating breath currents flow in the entire body through 72,000 identifiable nerve currents.

One of the chief reasons why we are unable to experience the mysterious process of death, is fear of death. Fear does not allow us to remain aloof – separate and meditate. Fear does not allow us to trust nature. This fear is the result of hostility to nature, one of the powerful and distinct manifestations of ego system. Only way out is wherever we have hatred or prejudice towards any living being, start replacing it with love or compassion. Start experimenting with trust in nature and analyse the result. You will find a living, responsive and friendly nature protecting our new instinct of love and compassion, helping us to come out of our negativity of dislike and prejudice.

When mystery of death is understood, yet not experienced, we will find that end of full life span leaves us with tired nerves, diseased body, sometimes irrational temper. Many of our friends and relatives of our age might have passed away and we might not fit well in the company of younger people. We might feel completely out of tune in this company. Mechanism of death therefore is a benevolent mechanism of nature to take us out of misery and monotony and puts us into a new environment.

If our karmic tract is good, after death we are born in a good and pious family and have 20 years of rest i.e. childhood and young age before new responsibility of adult life. Hence the process of death also leads us to rest and recuperation.

In every cycle from birth to death, a human being is developed to ultimately uncover the laws of nature concerning his birth, death, happiness and misery. However, in each natural cycle, progress is up to a certain limit and a new cycle is necessary for further progress. It is a step-wise ascent. In the presence of ego system which invariably is present, the journey is tiring. It requires rest on the way just like way-side inns during pilgrimage.

Before death, nature makes us unconscious, all the breath currents get involved like a seed of banyan tree and this invisible seed is then put in a fresh ground – a mother's womb, with all the necessary requirements like water, food, air etc. to take us on a new onward journey – a path divine.

24

Law of Retribution

"Reactions of all your actions would only give you reward or penalty. God has nothing to do with that."

~~

This law popularly known as karmic law deals with a simple fact in our worldly existence that whatever good or bad things we do, comes back to us in a similarity of experience wherein the event is not similar but the experience of an event is more or less similar.

Any human being under the effect of ego does mixed types of actions – some positive and some negative. Invariably, his or her present life is also a mixed event – if the business is all right, there may be health problems in the family or sometimes, family relations may be bitter and painful. At times, for a part of the life, there may be financial and family problems and part of it may be reasonably happy.

At times, a mixed pattern of few alternate years of comfort and misery slowly going down to a miserable pattern of existence is generated. When such a thing happens, we blame the people who have been responsible for such events. Really, these events are the result of our actions based on our arrogance, jealousy, prejudices, fanaticism etc. Blaming others or trying to correct an outward situation, we end up in a tight grip to our neck - with law of diminishing returns. We try to solve a problem and it gets confounded.

Presently, business and industry develop hostility to nature by polluting air, wasting water, exploiting earth – more fertilizers, more oil wells, destruction of forests, over exploitation of marine products, rearing animals for meat production, destroying wild life etc. Law of retribution operates by way of famine, war, epidemic, tidal wave, tsunami, earth-quake etc. Only way out is to stop exploiting nature rather than try to prevent natural calamity by new researches.

To some extent, nature's destructive cycle is a part of the development pattern. However, when things go in excess, then nature balances through natural calamity. When one community starts hating another community as in the case of Hindu-Muslim riots, then that hatred is brought down by a natural process of war whereby the futility and misery of war makes people weak, poor and leads them on spiritual path which works in the direction of reducing hatred.

In case of USA and Arab countries, sheer extravagance of a luxurious living results in broken families, addiction to drink, drugs etc. Here law of retribution results in incurable psychological diseases, AIDS, Cancer etc.; way out is treatment plus change of life style. Poverty in India is nature's punishment for offences on *Shudras* like *Harijans* and widows and exploitation of farmers and labourers.

Whenever we kill anybody, insult anyone or develop prejudice towards individuals, these negative impressions get stored in our brain. After a time, they attract such individuals who create situations whereby we are exploited, or uncared for or people harm us for no apparent fault on our part. Under these situations, evidence of revenge or hatred is present in our mind. If we accept these situations gracefully, they pass away leaving us purer. This is the teaching of the law of retribution. Nothing happens to us for which we are not responsible. Way out of our misery is the acceptance of difficulty and not fight outside.

Other corollary of this law is that as soon as we accept the difficulty, nature starts building strength in us to keep calm during such periods.

There is no penalty in nature except that which we deserve. That penalty is also the result of repeated ignorance of nature's warning that the path followed by us is wrong. Law of retribution is the infinite mercy of nature – because ultimately
it is the difficulty which puts us on the right path.

25

Law of Limits

"What is the essence of the world? Normality! Above Normal is the poison. Below Normal is the poison too. Come to Normality!!"

~~

Whatever changes we see in the world, they are basically of two types, either growing or decaying. This process takes place with increasing or decreasing speed and so also the process of decay. The process of growth and decay and its changing speed however reach a limit beyond which it neither grows nor decays. Transformations do take place, but growth and the speed of growing reaches a limiting value beyond which it reverses.

Understanding a natural law helps us to see through the workings of nature. In this case, knowing the law of limits, helps us to keep patience when we are subjected to an uncomfortable process like misery, sickness, etc., especially when we know that it reaches a limiting value after which it reverses. In some cases, when we die in a miserable condition, process of transformations takes place to create more favourable conditions for reversal. But the process of reaching a limiting value and then reversing does take place. Near the limit, process of growth or decay slows down so that we feel that we are steady – neither increasing nor decreasing. In fact, process of growth and decay is continuously on.

The law of limits also applies to growth of population, pollution, noise, fuel consumption etc. In case of population, the number is brought down by earth quake, tidal wave, war, epidemic etc. Nature has enough weapons in her hands to restore the balance once the limit is reached. When the rate at which population is increasing slows down, then process of reversal is nearing. This is the advantage of understanding a natural law.

In case of childhood, when it reaches the limiting value, it transforms itself into a youth. When youth reaches a limiting value, it transforms into old age and old age transforms at the limit to a new cycle of childhood through a discontinuous process of death and rebirth. Knowing the law allows us to co-operate with the workings of nature and not get too much attached to wife, children, money, etc. because similar things will also be available in the next cycle.

When wealth or happiness is increasing, knowing the law stops us from being excited because we know that such phenomena reach limiting values after which they reverse. So also regarding poverty and sickness, they reach a limiting value and reverse in the same cycle or through transformation to a new cycle, but no misery, poverty, sickness or unhappiness is permanent. Hence, we can cultivate patience and steadiness of mind.

Ninety percent of illnesses are self-terminating. Even if we do not treat, they reach a limiting value and reverse. As soon as some difficulty comes in the form of sickness, nature starts producing anti-bodies to combat the sickness and thereby speed of sickness slows down indicating approaching limit and transformation. A sickness, when it ends is not only a process of healing, but is also a process of transformation in nature's development path.

Alternate cycles of happiness and misery, health and sickness, poverty and prosperity with their alternating limits are a part of nature's development path with a purpose and meaning which is carefully hidden. It is not secret but carefully covered up, sufficient to create an illusion. Such is the way nature works. It requires some Copernicus to declare that it is not the sun which is moving around the earth, but the other way. Learning to read the codes-message from nature frees us from ego and its consequent misery. Law of limits is one such operation.

One of the message nature continually gives us is to warn us – remind us to closely look at the natural phenomena by living with nature and thus decode its methods of working. Mother nature is continuously inviting her divine child to come to her bosom; as soon as the child comes, mother's breast becomes full of milk of understanding which nourishes the child and fulfils the mother.

26

Sanctity of Wealth

"Intent counts. What is one's inner intention behind each act is important. If you have intent of using your skills and knowledge for helping others, money would automatically come as a by-product."

~~

Wealth is produced by nature as a result of human work. Purer the mind of a human being producing the wealth, it gets produced in abundance; whereas with an impure or greedy mind, wealth is not produced. For example, in case of agriculture due to famine or draught; or in other methods of manufacturing or trading, it produces losses due to family quarrels and disturbed mind.

Wealth gets attracted towards the mind which is full of compassion and love, mind which becomes serene after progressive extinction of ego.

Wealth when produced should be used up for our own expenses or for expansion of business or for social work. A part can be kept aside for sickness, children's education or some such work which is required for fulfilling house-holder's responsibility.

It should never be stored beyond necessary minimum because excess always escapes through any available unclean route.

It is possible to produce more wealth, use more wealth, distribute more wealth, but it is never possible to store more wealth. It is theoretically impossible.

One of the biggest mistake, a human being makes, is to make money out of money by giving it on interest and again by increasing the rate of interest to squeeze the borrower. Not to pay tax on the interest by taking a part of it in an unaccountable manner. To collect No.2 or black money by siphoning money by wrong purchase bills and to invest this money again on interest which is also unaccountable, and lastly to use only interest and keep the unaccounted money safe.

The purpose of investing in land, stocks, speculation or mutual funds is to earn money without work.

Prophet Mohammed was the first to declare that making money out of money is sin. It is prohibited in Islam. Next was Karl Marx who said the same thing.

Money which is earned by our own work and spent wisely will bring us peace and harmony.

Only certain time should be spent on earning, leaving enough time for family and leisure. There must be a limit that whatever is produced in eight hours of work should be sufficient or I must cut my expenses so that I stay within eight hours work.

Nature provides enough for our needs but never enough for our greed.

While earning we must be careful, to wish well of our competitor. Increase in our wealth, if at all necessary, should be by increasing our efficiency and organizing power.

Borrowing should be limited to twenty-five percent of our capital. There should be no money with us which cannot be disclosed to others.

While earning, maximum attention should be paid to keep our mind pure. There should not be boasting, superiority, unfair means, fraud, or showmanship. Respect for family members, relatives, staff members, servants should be one of the aims.

Test of the pure wealth would be that mind will be relaxed, free from disease and number of benevolent works will be undertaken by wealth.

More wealth should lead us to genuine Humility.

Economy is not the minimum use of money so that balance can be invested to earn maximum interest or return as in case of speculation, shares and mutual funds. Economy is the optimum use of money so that balance can be used for helping others who are in need of money.

Genuine wealth earned and optimally saved should lead us to an equanimous, peaceful and joyful life even in the midst of difficulty.

Loss of wealth leads us to strengthening of our weak links; at times, it leads us to spiritual path, cuts extravagant and unnecessary expenditure.

27

Human Relations

"One who has conquered his ego can please any person and resolve issues with equanimity."

~~

Complementary Nature of Human Relationships :
During our *Satsang* with Dada Bhagwan, one day we hit upon this subject of complementary nature of human relationships. In short, what he wanted to say was that whenever human beings come together, may be in small groups like family, neighbours or business partners, the people living together will have different temperaments, abilities and emotions. There is a spiritual law governing this relationship; the natural law states that, "a weak point of one in a relationship will be complemented by a strong point of the other and vice versa."

An untrained mind always looks at the weak or negative side of the other partner and thereby magnifies those qualities which result in friction, quarrel etc. However, if the mind is trained comprehending this law we can meditate on the positive side of the other and as a result we would cherish mutual love, harmony, peace and creativity.

Ego – The origin of Conflict in Relationships:

Everyone possesses ego in some measure. In present times, most of the people have strongly developed ego systems difficult to control. This ego is at times over-developed, expressed, or otherwise latent or suppressed. It leads to a strong sense of possession and emotional disturbances. It also leads to conflict, harsh words, selfish behaviour, erratic nature and consequent loss of peace, wealth and health.

If human beings are taught how to tolerate dominance of the other partner, to meditate how to develop the totality of acceptance, then the world would be divine. Tolerance does not mean suppression, but it means intelligent understanding and adjustment with the egoistic partner.

There is a law in nature that negative charge attracts positive charge and similar law is always operating when two people come together. Initially there are turmoils in any close relationship whether it be married couples, father-son, mother-in-law – daughter-in-law, two brothers or business partners etc. This is essentially due to the fact that one partner expects the other to have the same qualities as he or she has and cannot understand why this is not so. This results in unrealistic expectations and misunderstandings. Patience and tolerance over a period of time will lead to settling down with happy relationships.

If this law of complementary relationships is taught initially as a part of inter-personal relationship, considerable waste of energy can be saved.

Dealing with Children:
Husband-wife relationship when properly adjusted becomes a profound positive influence on children. The restlessness of children is converted into peaceful constructive energy and they become better citizens of their country.

Dealing with Old and Sick:
There will be no family where there will be no old and sick people and one has to learn how to deal with them. Generally old people have plenty of wisdom and maturity and in these days when both husband and wife are working, the old people contribute considerably in looking after home and the children; but a time comes when they also become sick, sometimes develop an erratic temperament and one has to understand that this is the time when one needs to practice enhanced equanimity while serving the parents.

Discipline:
Wherever human beings work together in a family or a business office, discipline becomes unavoidable since without it there might be chaos. But discipline cannot be enforced, we have to live a self-disciplined life ourselves and slowly persuade others to do so. At times, one has to be harsh without being bitter. Understanding and practice of this law acts as a lubricant in enforcing discipline in family and office.

Controlling Impulse:
For practicing this law of complementary relations, it is necessary to control the impulse of anger, jealousy, strong possessiveness

etc. The mind also should be tamed not to react with adverse situation but to change it with love, perseverance and humility. Introspection of one's own behaviour will help in controlling impulse. Ultimately the law of love will prevail.

Positive Thinking:
Left to itself, the mind always tends to observe and think what is wrong with the society, with the Government, the locality, neighbours and family members. If we look closely, there are so many positive things happening – widening of roads, increasing of water supply, accommodations, jobs, infrastructure, cleanliness, education, rural development and so on. The mind which is trained to think positively will be able to practice law of complementary relations easily and set an example for many others to change their lives.

Humility:
J. Krishnamurthy has said that humility is not social etiquette. Usually people practice humility by being nice to other people in talking, sometimes during a party, dinner, looking after their minor needs or talking pleasantly on telephone etc. They remain in illusion that they are decent human people. However, humility is to be practiced at a deeper level of consciousness and consists of serenity and genuine love towards all those with whom we deal. It also has an attitude to share with others whatever he or she has. This quality also requires training of the mind.

Evolution:
A human being can be considered evolved only when he or she can control his temper under adverse situations, resolve problems and situations amicably, is able to let go many things which are called personal and only use the tool of love, cheerfulness, equanimity, forgiveness, adjustment and magnanimity in dealing

with others.

Widening the Consciousness:

Once we have learnt this law of complementary relationships and consequent adjustments then it would be easy to deal with a larger group say children, parents, neighbours, partners etc. We can apply this law to large number of people working together anywhere. Ultimately, this leads to the widening of our consciousness to infinity.

Appendix

1. Qualities of central path in spiritual evolution with deviations which need to be corrected

	PLUS	CENTRE	MINUS
1	Attachment	Love	Hatred, Prejudice
2	Emotion	Compassion	Cruelty
3	Excitement, Restless	Cheerfulness	Morse, Sad, Gloomy, Negative
4	Inactivity	Equanimity	Short-tempered
5	Substanceless, Hypocrisy	Humility	Arrogance
6	Recklessness	Fearlessness	Timidity
7	Defenceless	Straight forwardness	Crookedness
8	Flattery	Sweetness	Bitterness
9	Rigidity	Firmness	Looseness
10	Athletic, Over-importance to physical body	Good health	Disease
11	Extravagance	Prosperity	Poverty
12	Looseness	Forgiveness	Revenge
13	Dullness	Patience	Restlessness
14	Blind action	Perseverance	Inertia: Leaving the aim with minor difficulty

	PLUS	CENTRE	MINUS
15	Lethargy	Serenity, Peace Rhythm & harmony	Disturbed, Turmoil
16	Glamour, Excess etiquette	Beauty	Harshness
17	Obsessive compulsion for cleanliness	Cleanliness	Dirt
18	Craving	Intense activity without restlessness	Inactivity
19	Over dependence, Subserivent attitude	Gratitude	Ingratitude, Arrogance
20	Superiority complex	Name & fame	Ignonimity
21	Missing/ Overlooking the totality	Penetrating insight	Uncontrolled behaviour
22	Exciting people	Motivation	Herd mentality
23	Intellectual, Argumentative	Smartness, Sharpness	Dull-gross mind
24	Overactive	Energetic	Lethargic
25	Extravagant	Charitable	Miser

2. Outcome of Negative Emotions

	Negative Emotions	Leads to	Resulting in
1	Hatred	Strong dislike	Imposition of one's will
2	Greed	Lust	Disease
3	Arrogance	Bitterness	–
4	Expectation	Strong likes	Strong likes
5	Jealousy	Strong ambition	Loss of temper
6	Anger	Restlessness (Hyper-activity)	Harsh words
7	Revenge, Animosity	Aggression	Fault finding nature
8	Fear	Frustration	Loss of tolerance
9	Obstinacy, Hypocrisy	Craving	Heart disease: Strong ambition
10	Strong sense of possession	–	Lack of control over desire
11	Over passionate	Fraud or cheating	Cancer: Bitterness, Arrogance
12	Attachment	Prejudice	Diabetes: Anxiety, Neurosis

3. Outcome of Positive Emotions

Positive Emotions	Leads to	Resulting in
1 Peace	Open mind	Morality and ethics
2 Patience	To observe truth	Discriminating intelligence
3 Perseverance	Proper choice	Sweetness in speech
4 Acceptance	Charity	Straightforwardness
5 Love	Truthfulness	Sharp defence
6 Compassion	Self-confidence	Forgiveness
7 Cheerfulness	Equanimity, Firmness	Self-control

About the Author

Jayantilal Shah, fondly known as Kaka was born in a modest Jain family on 2nd Sept 1925 in Dhoraji, a small town in Gujarat. After doing his schooling there, he moved to Rajkot to do his Intermediate at Dharmendra Sinhji College. Near the college there was the Ramkrisna Math, which soon became his spiritual college. It was here that he got introduced to Swami Vivekanand's writings, which had a great impact on him at this early age. For further studies he came to Mumbai in 1943 and joined the Royal Institute of Science while staying at Mahavir Jain Vidyalaya Boarding.

As he continued his studies, he became strongly influenced by Mahatma Gandhiji's freedom struggle and philosophy, which remained with him throughout his life. During these formative years, he also developed a deep interest in the Bhagwad Gita, which guided him through all his future endeavours. Adding several such interests along his journey, he graduated as a brilliant Chemical Engineer in 1947 and worked with Balmer & Lawrie as a Chief Chemist for eleven years.

He progressed well in his professional career and at the same time looked after his extended family, but devoted considerable time for his spiritual growth and for contributing towards society. Sharing his wealth with the needy came naturally to him. Continuing this journey, the teachings of Vinoba Bhave and Srimad Rajchandra also highly influenced his thinking and guided him in his spiritual path. He made sure to find the time to meet and help individuals and groups working with the similar kind of ideology. Having varied interests, he would read extensively on various subjects – religion, spirituality, Ayurveda, history, science, the writings of various thinkers like Tagore etc. He also developed a keen interest in music and was an avid traveller.

In 1973, he started his own industry, Anand Engineers, and during that same period he was introduced to *Vipassana* meditation as taught by Shri S. N. Goenkaji. He also came in contact with Shri Dada Bhagwan who became his living guru, and to whom he remained indebted all his life. His Company, Anand Engineers, was run inculcating the divine values as learned from Shri Dada Bhagawan, as well as, the practice of *Vipassana* meditation. Soon he developed a model – Applied spirituality in business and tried it out in his own company – with great success. Satisfied with his spiritual experiment, leaving the enterprise with his children and partners, he retired in the year 2005 but his journey continued…

Soon after, his house became an ashram as it drew people from all walks of life who would come with their doubts, problems, difficulties be it in their personal life, family, business, society, etc. He would listen, counsel and help to change their lives for the better. Many benefited enormously from his intervention. Till the age of 91, he would happily meet and guide them by his deep knowledge of Akram Vignan, his faith in the *Bhagavad Gita* and the practice of *Vipassana*.

He had developed a unique approach of guiding people, which could perhaps be termed as "Kakaism".

On 1st July 2016, he undertook a different journey, but left behind many who are forever indebted and full of gratitude for what he did for them.

www.ingramcontent.com/pod-product-compliance
Lightning Source LLC
Chambersburg PA
CBHW071033240526
45469CB00006BD/2196